Defending the Arsenal

One important area of interest within military and policy circles focuses on an effort to revitalize the nuclear triad amidst a number of competing strategic interests. The difficulties arising from US engagement in Iraq and Afghanistan are leading many scholars and policy makers to question whether a reinvigorated nuclear triad has any role in deterring modern adversaries. This volume takes an unashamed pro-nuclear modernization position and argues for designing and fielding new nuclear warheads and delivery systems (submarine, ICBM, and bomber) while also arguing against signing the Comprehensive Test Ban Treaty or agreeing to further reductions in the nuclear arsenal. It also argues that nuclear deterrence remains as relevant today, perhaps more, than it was during the Cold War. With so many authors advocating for "Global Zero" and highlighting perceived dangers from a nuclear arsenal, this work stands in stark contrast to the chorus of anti-arsenal works. Because of the work's structure and effort to answer questions of current relevance, it should appeal to a broad audience including: service staffs, PME students, COCOM staffs, Pentagon personnel, Capitol Hill staffers, policy makers, academics, graduate students, and interested readers.

Adam B. Lowther, Director, School of Advanced Nuclear Deterrence Studies.

Stephen J. Cimbala, Penn State University–Brandywine.

Defending the Arsenal

Why America's nuclear modernization still matters

Edited by Adam B. Lowther and Stephen J. Cimbala

Routledge
Taylor & Francis Group

LONDON AND NEW YORK

First published 2017 by Routledge

2 Park Square, Milton Park, Abingdon, Oxon OX14 4RN

605 Third Avenue, New York, NY 10017

Routledge is an imprint of the Taylor & Francis Group, an informa business

First issued in paperback 2021

British Library Cataloguing-in-Publication Data
A catalogue record for this book is available from the British Library

Library of Congress Cataloging-in-Publication Data
Names: Lowther, Adam, editor of compilation. | Cimbala, Stephen J., editor of compilation.
Title: Defending the arsenal : why America's new arsenal still matters / edited by Adam Lowther and Stephen Cimbala.
Description: Milton Park, Abingdon, Oxon ; New York, NY : Routledge, 2016.
Identifiers: LCCN 2016015761| ISBN 9781138204546 (hardback) | ISBN 9781315460697 (ebook)
Subjects: LCSH: Nuclear weapons–Government policy–United States. | Arsenals–United States. | United States–Military policy. | Deterrence (Strategy) | National security–United States.
Classification: LCC U264.3 .D44 2016 | DDC 355.02/170973–dc23
LC record available at https://lccn.loc.gov/2016015761

ISBN: 978-1-138-20454-6 (hbk)
ISBN: 978-1-03-217942-1 (pbk)
DOI: 10.4324/9781315460697

Typeset in Times New Roman
by Wearset Ltd, Boldon, Tyne and Wear

Contents

Figures

Tables

Contributors

Stephen J. Cimbala is Distinguished Professor of Political Science at Penn State University–Brandywine. Dr. Cimbala is the author of numerous works in the fields of nuclear arms control and US national security policy, among other topics. A past winner of Penn State's Eisenhower Award for Distinguished Teaching, Dr. Cimbala has recently authored *The New Nuclear Disorder* (Ashgate, 2015) and co-authored *US National Security* (with Sam C. Sarkesian and John Allen Williams, Lynne Rienner, 2013).

Michaela Dodge is a senior policy analyst at The Heritage Foundation, a Washington, DC-based think tank, where she works on nuclear weapons and missile defense policy issues. She holds a master of science degree from Missouri State University and is a doctoral student in political science at George Mason University.

Anita Feugate-Opperman is the Director, Commander's Action Group for Air Force Global Strike Command. She received her commission through the Air Force Reserve Officer Training Corps in May 1994. She holds a bachelor of arts degree in political science from Indiana University, a master's degree in information systems management from Lesley University, a master's degree in military operational art and science from Air University, and a master's degree in military strategy from Air University.

Robert A. Hoskins is the Personnel and Training Panel Chair in the Air Force Corporate Structure. He began his military career as a personnel officer in August 1995, and earned his commission from the USAF Officer Training School. He holds a bachelor of science degree from Indiana University, and master's degrees in military operational art and science and national security studies from Air University.

Shelley Bischoff Kavlick is the acting vice commandant of the Officer Training School (OTS), Maxwell Air Force Base, Alabama. OTS is the largest Air Force annual accession source, training over 2,600 total-force officers for the Active, Guard, and Reserve components. Col. Kavlick graduated from the University of California at Davis and was commissioned a distinguished graduate from the Air Force Reserve Officer Training Corps in

1995. She holds a master's degree from the Royal Military College of Science, Cranfield University, England. Col Kavlick started her career as an aircraft maintainer.

Thomas C. Kirkham is chief, Air Force Weapons Safety, Air Force Safety Center, Kirtland AFB, NM. He enlisted in the Air Force Reserves in April 1987 and was a crew chief on the C-5A Galaxy. He was then commissioned through the Reserve Officer Training Corps in 1993 and has since served in a variety of maintenance and logistics assignments at squadron, wing, major command, and unified command levels. He is a master aircraft/munitions maintenance officer with experience on the B-2A, B-52H, F-15E, and MQ-1 aircraft as well as numerous tactical nuclear and conventional weapons. Col. Kirkham is a graduate of the Joint Forces Staff College and has a bachelor of science degree in business management from Texas State University, a master of aeronautical science degree from Embry-Riddle Aeronautical University, a master of military arts and sciences degree from the Air Command and Staff College, and a master of strategic studies degree from the Air War College.

Adam B. Lowther (BA, Arizona State University; MA, Arizona State University; PhD, University of Alabama) is Director of the School of Advanced Nuclear Deterrence Studies at Kirtland Air Force Base. Previously, he was Research Professor and Director of the Center for Academic and Professional Journals at the Air Force Research Institute (AFRI), Maxwell AFB, AL. His principal research interests include deterrence, airpower diplomacy, and the Asia-Pacific. He is author or editor of five books and has published in the *New York Times*, *Boston Globe*, *Joint Forces Quarterly*, *Strategic Studies Quarterly*, and a variety of other journals and outlets. Dr. Lowther has served on the faculty of two universities where he taught courses in international relations, political economy, security studies, and comparative politics. Early in his career he served in the US Navy aboard the USS *Ramage* (DDG-61). He also spent time at CINCUSNAVEUR-London and with Naval Marine Construction Battalion 17.

Karyn E. McKinney is the squadron commander of the 381st Training Squadron. She began her military career as an enlisted medic in the Army Reserves in 1987, and earned her commission in the Air Force through the Health Professions Scholarship Program in 1998. She has a doctor of medical dentistry degree from the University of Pittsburgh, a master in public administration from Troy State University, a certificate of training for the Two-Year Advanced Education in Dentistry Program from Wilford Hall Medical Center, and a master in strategic studies from the Air University.

Eric Y. Moore is the Air Force Global Strike Command chief of ICBM Maintenance and is responsible for all ICBM sustainment at locations throughout Wyoming, Nebraska, and Colorado in direct support of United States Strategic Command and the National Command Authority. His career spans ICBM maintenance, munitions maintenance, and the acquisitions fields. Col.

Moore was commissioned from the Air Force Academy in 1992 and is a graduate of the Air War College.

Donald M. Neff enlisted in the Nebraska Air National Guard in February 1989 and served as an RF-4C Photo Sensor Maintenance Specialist in the 155th Tactical Reconnaissance Wing. He graduated from the University of Nebraska–Lincoln in 1993 and received a commission through the Reserve Officer Training Corps. He began his military flying career assigned to the 173rd Air Refueling Squadron as a copilot and later aircraft commander in the KC-135R Stratotanker. In 1998, Col. Neff transferred to the 326th Airlift Squadron, Dover AFB, DE, and flew the C-5A/B Galaxy. He returned to the NEANG and the 155th Air Refueling Wing in 2001 and has served in several capacities including squadron and group commander. Col. Neff has combat and combat support flight time in operations Allied Force, Southern Watch, Northern Watch, Noble Eagle, Enduring Freedom, Iraqi Freedom, New Dawn, Unified Protector, and Inherent Resolve. A pilot for United Airlines since 1999, he has flown the Boeing 727, 737, 757, 767, and Airbus A-319/320.

Paul D. Schumacher is assigned to the Office of the Assistant Secretary of Defense for Strategy, Plans, and Capabilities serving as the Deputy Director, Nuclear Deterrence Policy. Col. Schumacher is a Reserved Officer Training Corps distinguished military graduate from Mercer University, Macon, GA. He was commissioned a Regular Army, Second Lieutenant in the Field Artillery in 1991. He served in a variety of command and staff positions from battery to brigade level in both towed and rocket artillery in multiple CONUS locations. He transitioned to Functional Area 52, Nuclear Operations and Counter-proliferation, in 2004 serving as a nuclear policy officer in US European Command and assistant professor of chemistry, United States Military Academy. He holds a bachelor's degree in chemistry from Mercer University, and has a master's degree in neuroscience and a doctoral degree in radio-chemistry from Washington State University. He is a graduate of the Field Artillery Officer Basic Course and Advanced Course, Combined Arms Service Staff School, Command and General Staff College, and the Air War College.

Michelle K. Stinson is the program manager for Foreign Language Mission Integration at the Office of the Director of National Intelligence. She began her career as an Army military intelligence and foreign area officer, serving primarily in overseas assignments in Europe, the Middle East, and Asia. She continued to work in positions supporting the federal government after retiring from the Army. Ms. Stinson holds a Bachelor of Arts degree from Auburn University, a master's degree in Strategic Studies from the Air University, and a Master of Arts degree from the Johns Hopkins School of Advanced International Studies (SAIS).

Foreword

By the time World War I ended in 1918, over 16 million people died worldwide due to the war. The world talked about the war to end all wars. Yet, less than a generation later World War II began, and conservative estimates suggest over 68 million people died as a result of the war (with 100 million total casualties). History is replete with examples of the times when conventional deterrence failed, and there are memorials all over the globe commemorating those times. However, nuclear weapons came along in 1945, and for the past 70 years, these weapons have been instrumental in deterring major conflict between world powers.

Occasionally, misinformation circulates in the public that hypes the irrelevance of nuclear weapons in the current world environment. This misinformation continues to be echoed until someone like Russian President Vladimir Putin or North Korean leader Kim Jong Un makes outrageous saber-rattling statements mentioning the use of nuclear weapons. Then suddenly the world's press, academia, policy makers, and the public are reminded that nuclear weapons are still used as a tool for deterrence.

Since the end of the Cold War, the possibility of nuclear war has become more remote; however, by no means does this mean nuclear conflict will not occur. Some would argue that today we have entered a new nuclear age—characterized by more complexity, more players, less predictability, and different rules ... all of which make deterrence harder. To continue deterring conflict, the US arsenal must be credible in the eyes of our adversaries. However, today's TRIAD of nuclear forces ... submarines, ICBMs, bombers, and weapons are well past their design life. The takeaway is the US needs to rebuild our deterrent strength and modernize its nuclear force. Our potential adversaries understand strength and any weakness or indecision on our part factors into their risk/reward calculus.

The stability the US nuclear arsenal brings to our country needs to be better communicated to the public. This stability is not only important for the deterrence of conventional and nuclear wars, but also for the assurance it brings our allies. By assuring our allies, we allow them the freedom to act without fear or coercion from other nuclear states, furthering the aims of nonproliferation while at the same time bringing stability to the world stage to ensure World War III never happens.

Besides modernizing our nuclear TRIAD systems, we need to modernize our deterrence thinking for the twenty-first century. We need to ensure that our doctrine and policy are updated to match the complicated current world environment. By educating ourselves and the public on nuclear deterrence theory and pressing issues such as escalation control, extended deterrence, and nonproliferation of nuclear weapons we can make a more credible nuclear deterrent and raise the threshold for nuclear use.

Recent events in Russia, North Korea, Iran, and China have made this an ideal time for a book that will help to provide a better understanding about nuclear weapons and deterrence in the twenty-first century. In order to match our deterrence to the evolving world environment, we need the best and brightest minds to offer their ideas and opinions. This book does a great job of pulling from various viewpoints and topics from students enrolled in the Nuclear Issues Elective as part of the Air War College Strategic Studies Master Degree Program.

Defending the Arsenal: why America's nuclear modernization still matters not only offers a multi-faceted approach to the importance of deterrence and nuclear weapons, but it pulls from several different perspectives including active duty military, reserve component, and military retirees as well as civilian personnel and academia. Each chapter offers a unique viewpoint of nuclear deterrence, helps to paint a vivid picture of the importance of our nuclear arsenal, and reminds adversaries that the consequences of attacking the US or her allies far outweigh any potential benefit.

In today's complex world, it is important to continuously evolve our thinking to ensure our deterrence concepts and readiness remain credible to meet the demands of the twenty-first century. With the ideas and concepts offered in this book, I am confident the US will be ready to meet these demands.

Stephen W. Wilson
Lieutenant General, USAF
Commander, Air Force Global Strike Command

Acknowledgments

Grateful acknowledgment is made to the following organizations and persons for their interest in, and support for, this research. At Penn State University–Brandywine, we appreciate the administrative support of Dr. Kristin Woolever, Chancellor; Dr. Cynthia Lightfoot, Director of Academic Affairs; Professor Susan Ware, head librarian; and Justin DiMatteo, head, Information Technology Department; and their respective staffs. A research grant from the Department of Political Science, College of the Liberal Arts, was also helpful.

We would also like to thank the Air Force Research Institute and Lt. Gen. Al Peck (Ret.) and Dr. Dale Hayden for support of this research project. The time needed to prepare this manuscript was of great utility.

A particular debt of gratitude is owed to Jessica Lowther, who volunteered to edit the manuscript, fixing all of the many mistakes and incorrect notes. Her time and effort in preparing the manuscript for the publisher are greatly appreciated.

Our intellectual debts are even greater than our institutional ones. We are especially grateful for pertinent subject matter insights and general scholarly encouragement to: Stephen Blank, Paul Bracken, Daniel Cohen, Paul Davis, Andrew Futter, David Glantz, Colin Gray, Michael Guillot, Dale Herspring, Hunter Hustus, Jacob Kipp, Lawrence Korb, Keith Payne, the late Sam Sarkesian, James Scouras, Michael Tate, James Tritten, and John Allen Williams. None of these persons is responsible for any arguments or opinions in this work.

Finally, it has been a pleasure to work with the excellent staff at Taylor & Francis. This book would not have been possible without their professional commitment, hard work, and diligent follow-through on all aspects of the project.

Abbreviations

A2/AD	anti-access, aerial denial
ABM	anti-ballistic missile, also BMD for ballistic missile defense
AEGIS	missile defense system
ALCM	air-launched cruise missile
ASAT	anti-satellite weapon
ASBM	anti-ship ballistic missile
B-2	heavy bomber
BMD	ballistic missile defense, also ABM for anti-ballistic missile
C3	command, control, and communication
C3I	command, control, communications, and intelligence
C4ISR	command, control, communications, computers, intelligence, surveillance, and reconnaissance
CBM	conventional ballistic missile
CPGS	conventional prompt global strike
CTBT	Comprehensive Nuclear Test Ban Treaty
DIA	Defense Intelligence Agency
ELINT	electronic intelligence
EMP	electromagnetic pulse
GPS	global positioning system, satellite-based
ICBM	intercontinental ballistic missile
IMS	International Monitoring System
JCS	Joint Chiefs of Staff
LEP	Life Extension Program
LRS-B	long-range strike – bomber
MAD	mutually assured destruction
Minuteman	intercontinental ballistic missile
MIRV	multiple independently targetable reentry vehicle
MRBM	medium-range ballistic missile
NATO	North Atlantic Treaty Organization
New START	US–Russian nuclear arms reduction agreement (Strategic Arms Reduction Treaty)
NNSA	National Nuclear Security Administration
NORAD	North American Aerospace Defense Command

NPR	Nuclear Posture Review
NPT	Nuclear Non-Proliferation Treaty
NSA	National Security Agency
NSC	National Security Council
PGS	precision global strike
SIGINT	signals intelligence
SLBM	submarine-launched ballistic missile
SRBM	short-range ballistic missile
SSBN	strategic ballistic missile submarine
SSGN	cruise missile firing submarine
SSP	Stockpile Stewardship Program
STRATCOM	Strategic Command
Trident	missile and ballistic missile submarine

Introduction

Stephen J. Cimbala and Adam B. Lowther

Modernization of the US nuclear arsenal is one of the most urgent questions of policy and strategy in the national security environment of the twenty-first century. The need for modernization of US strategic nuclear forces is not only about fielding capability and budgets, or roles and missions, although these are important. Strategic nuclear force modernization also impacts upon the ability of the United States to exert appropriate international leadership, including deterrence of adversaries, assurance of allies, and the preservation of deterrence and arms race stability. The preservation of nuclear deterrence and arms race stability rests on a paradox. Once fired in anger, nuclear weapons would cause unprecedented destruction in a short time. The lethality of nuclear weapons, however, gives them their suasion as instruments of deterrence.

The end of the Cold War and the demise of the Soviet Union ushered in new requirements for nuclear deterrence, tailored to a wider post-Cold War environment of possible threats. Instead of a singular US and NATO focus on deterring the former Soviet Union, the US now requires forces with the flexibility to deal with threats from rogue states outside Europe, from non-state actors, and from "failed states" in possession of nuclear weapons or fissile materials. The US strategic nuclear deterrent, including land-based intercontinental ballistic missiles (ICBMs), submarine-launched ballistic missiles (SLBMs), and long-range heavy bombers, remains as singularly important as it was in the previous century. It provides the top cover of military capability that protects the United States and its allies against nuclear attack or blackmail, and it enables the US as a leader in the field of nuclear arms control and nonproliferation.

The United States takes seriously its responsibility for leadership in nuclear arms control. President Barack Obama acknowledged at the outset of his administration that his further objective of nuclear abolition would probably not be achieved in his lifetime, if ever. Therefore, according to his nuclear policy guidance, the US must maintain a nuclear deterrent that is second to none as long as nuclear weapons exist and are deployed by other states. Nevertheless, the Obama administration has sought to diminish the US dependency on nuclear weapons for dealing with non-nuclear threats. This objective must be tempered by awareness that nuclear and non-nuclear forces have certain overlapping support capabilities and taskings (for example, the US bomber force carries out both nuclear and conventional long-range missions). In addition, both nuclear and conventional forces depend on cyber capabilities that are becoming increasingly important for deterrence and defense missions.

Even among those who favor modernization of the US strategic nuclear triad, there is some disagreement about the preferred mix of weapons systems and supporting capabilities necessary to fulfill US policy and mission requirements. For example, some argue that the US no longer needs three different "legs" or kinds of intercontinental delivery systems. One or more of the legs of the triad could be dispensed with, saving costs for modernization without actually jeopardizing US security. Others contend, to the contrary, that the strategic nuclear triad of land- and sea-based missiles and long-range bombers create a benign redundancy that helps to forestall prospective attackers. Any attacker would have to assume his ability to neutralize all three components of the US retaliatory force in order to limit damage to "acceptable" levels. Such an attack would be unlikely because each of the three legs of the nuclear triad has unique properties: ICBMs are most ready for prompt launch; SLBMs are the most first strike survivable platforms; and bombers allow policymakers the flexibility to send signals of nuclear resolve without making an irrevocable commitment to nuclear first strike or retaliation.

The future credibility of US nuclear forces will depend in part on the supporting infrastructure, including nuclear weapons laboratories, contractors, and idea factories inside and outside government. In this regard, there was much neglect during the first two decades after the end of the Cold War. Secretary of Defense Robert Gates's dismissal of the Air Force Chief of Staff and the Secretary of the Air Force in 2007, following an embarrassing lapse in security protocols for the transport of nuclear weapons, was a signature moment of revelation about a collective government and societal nuclear neglect. The "action" was in asymmetrical conflicts, including terrorism and unconventional warfare: government and think tank analysis followed the yellow brick road that bypassed the continuing need for nuclear weapons modernization and nuclear-related scientific knowhow. As a result, there is now a requirement, not only for a "linear" modernization of nuclear weapons and associated infrastructure, but for a broader based assessment of the role of nuclear forces in relation to other and emerging capabilities. For example, the development of Prompt Global Strike systems for long-range delivery of precision conventional weapons, including by platforms formerly dedicated exclusively to nuclear forces, creates a new offensive mix of capabilities relative to potential missions. Along with this, further research and development as well as deployment of US and allied NATO ballistic missile defenses (BMD) must factor into the assessment of US capabilities relative to those of other states. Third, the cyber realm will impose a discipline and a chaos of its own on nuclear modernization and force planning. Future planning scenarios will have to take into account the possibility of cyber-attacks either prior to an outbreak of kinetic war or during it.

National military strength and the effectiveness of preferred US military strategies are ultimately dependent on the support of the American people. In this regard, civilian policymakers, military leaders, and academics, among others, have a responsibility to provide research and analysis on important public policy issues related to nuclear weapons, including nuclear force modernization, deterrence, arms control, and nonproliferation. The present volume brings together academic and military collaborators in order to address these issues within the context of US Air Force current and future requirements and challenges. As the arm of service responsible for two of the three component delivery systems of the US strategic nuclear triad, the Air Force must operationalize and implement the wishes of policymakers, according to prevailing defense guidance and by using professional military judgment. This study is offered as a modest contribution toward the creation of an optimal policy-strategy linkage with respect to future US strategic nuclear forces and their assigned missions of deterrence and dissuasion of adversaries, reassurance of allies and others, and vital supports for nuclear arms control and cooperative threat reduction among nations.

1   The future US nuclear strategic environment

Michelle K. Stinson

Introduction

In the years ahead, the United States will confront a wide array of security challenges that include preserving strategic stability with a nuclear peer competitor, deterring the use of nuclear weapons by rogue nations, dissuading smaller nuclear powers from nuclear arms racing, preventing non-nuclear states from crossing the nuclear threshold, and preventing nuclear terrorism. These trends underscore the complexities of nuclear strategy, doctrine, and force structure, and support the premise that the United States must make quantitative and qualitative adjustments to its nuclear triad and current deterrence strategies to counter multiple nuclear threats in the strategic environment of the future. The American strategic arsenal of the twenty-first century must maintain strategic stability with Russia and China, deter potential regional adversaries, and assure allies and partners under the American nuclear umbrella.

Because of its seven decade record of success, American allies protected by the nuclear umbrella have chosen not to develop their own nuclear arsenal, a point too many elected officials and scholars take for granted. Nuclear abolitionists and other groups are calling for reductions in the nuclear arsenal and a new commitment to a world without nuclear weapons. Other groups have proposed significant reductions in the number of nuclear weapons or reductions in the number and mix of nuclear delivery systems.

This chapter will assess the future strategic environment from the perspective of US nuclear weapons policy, deterrence policy, the cost and structure of the nation's deterrent capability, and the future threat environment in order to discuss the quantitative and qualitative adjustments that are required for the nuclear triad and current deterrence strategies to counter multiple nuclear threats in the future. As long as these nuclear threats endure, the United States must have a strong nuclear deterrent that is safe, secure, and effective in meeting its security needs and those of its allies.

Does the United States need nuclear weapons?

The US nuclear deterrent has remained in continuous use every day since 1945 to ensure that an attack against the US or its allies would be unthinkable, given the devastating nuclear response that would follow. Many former senior policy-makers are leading the call for a commitment to a world without nuclear weapons.[1] Such a policy would leave the United States at a distinct disadvantage against nuclear competitors who are modernizing and expanding their arsenals. If the US were to move ahead with drastic nuclear reductions, it could lack the means to deter these advanced systems or provide a credible nuclear umbrella to allies and friends. Drastic reductions could also prompt a new nuclear arms race with states that seek to obtain nuclear parity with the United States.

Proponents of cuts to the US nuclear arsenal generally support three possibilities for decreasing or eliminating the nuclear forces: decreasing the size of the nuclear arsenal through reductions in warheads, but preserving the three delivery systems that make up the nuclear triad; eliminating one or more nuclear delivery systems; and/or deferring or canceling nuclear modernization programs.[2]

A new direction for US nuclear policy

The past four years have been historic in terms of setting a new direction for US nuclear strategy, policy, force posture, and funding. Proponents of cuts to the nuclear arsenal and nuclear abolitionists mistakenly declared victory in April 2009 when President Barack Obama pledged to pursue a world free of nuclear weapons during his "Prague speech."[3] While the speech did signal a new direction for the United States' nuclear enterprise and nuclear policy, the President also announced that as long as nuclear weapons exist, the United States must "maintain a safe, secure, and effective nuclear arsenal to deter any adversary, and guarantee that defense to our allies."[4] President Obama has acknowledged that a world free of nuclear weapons is a long-term goal that would not be realized quickly. The speech prompted an examination of US nuclear strategy, policy, and force posture which resulted in renewed support and funding for the US nuclear enterprise and the nuclear triad, while addressing the concerns of nuclear abolitionists with a new bilateral reduction agreement with Russia, a commitment to further cuts, and a pledge to eventually diminish the role of nuclear weapons in national security strategy.

With the Prague speech as guidance, the Department of Defense (DoD) led an interagency review to determine the future role of nuclear weapons and nuclear policy to include declaratory policy, acquisition, deployment, and employment, resulting in the 2010 Nuclear Posture Review (NPR) Report that outlined five key priorities:[5]

1   Prevent nuclear proliferation and nuclear terrorism;
2   Reduce the role of nuclear weapons;
3   Maintain effective strategic deterrence and stability at lower nuclear force levels;

4   Strengthen regional deterrence and reassurance of US allies and partners;
5   Sustain a safe, secure, and effective nuclear arsenal.

The NPR confirmed the fundamental role of the nuclear force in national security and updated declaratory policy by pledging that the United States will not retaliate with nuclear weapons against any non-nuclear state that abides by its Nuclear Non-Proliferation Treaty (NPT) commitments, relying instead on the threat of conventional military retaliation and ballistic missile defense capabilities to deter or defend against an attack.[6] Additionally, the NPR confirmed that while the United States will not develop new nuclear weapons to replace those in the existing arsenal, it will retain a smaller nuclear triad of upgraded intercontinental ballistic missiles (ICBMs), submarine-launched ballistic missiles (SLBMs), and heavy nuclear-capable bombers with modernized warheads and bombs in order to "maintain strategic stability at reasonable cost, while hedging against potential technical problems or vulnerabilities."[7]

In April 2010, the United States and the Russian Federation signed the New START (Strategic Arms Reduction Treaty), mandating that both countries will limit their nuclear weapons to a maximum of 1,550 operationally deployed strategic nuclear warheads on no more than 700 deployed ICBMs, SLBMs, and heavy bombers by February 5, 2018.[8] During the Senate's advice and consent process for ratification of New START, the Obama administration agreed to request more than $214 billion through 2020 to maintain, replace, and upgrade the nuclear force and nuclear weapons complex, ending a long hiatus in weapon modernization and delivery system upgrades, in order to support the NPR policy.[9]

In June 2013, the President announced the US Nuclear Weapons Employment Strategy to translate the conclusions of the 2010 NPR into more detailed guidance on the role and structure of nuclear forces for DoD planners.[10] While confirming the key objectives in the 2010 NPR, the strategy also includes the goal of eventually pursuing up to a one-third reduction in deployed strategic nuclear weapons from the level established in the New START Treaty to just over 1,000 nuclear weapons, while avoiding any discussion of nonstrategic weapons forward-deployed in Europe in support of NATO.[11]

Shortly after the White House released the new strategy, DoD submitted the Report on Nuclear Employment Strategy of the United States specified in Section 491 of 10 U.S.C. to Congress on behalf of the President.[12] The report assessed what changes to nuclear employment strategy could best support the five key objectives of nuclear weapons policies and posture outlined in the 2010 NPR and added a sixth objective: achieve American and allied objectives if deterrence fails.[13]

The new Nuclear Employment Strategy has disappointed nuclear abolitionists and advocates of Global Zero, although it does include support for moderate steps to reduce the numbers and role of nuclear weapons. The strategy reaffirms the requirement for counterforce options in nuclear planning and retains the nuclear triad. It also retains the capability to launch nuclear forces under attack;

continues the current alert posture; retains strike options against conventional, chemical, and biological weapons; confirms support for nonstrategic nuclear weapons in Europe; supports the storage and maintenance of a hedge of nondeployed warheads; and endorses the production of modified interoperable warheads.[14]

A new direction for deterrence policy

Effective deterrence in the future will continue to depend both on superior nuclear capabilities and the perception of a national will to respond to aggression with both nuclear and conventional weapons. Its practice will be complicated by the myriad of state and non-state actors that must be considered in developing effective and tailored deterrence strategies.

But the fundamentals of deterrence have not changed: *Joint Publication 1–02, Department of Defense Dictionary of Military and Associated Terms*, defines deterrence as the "prevention of action by the existence of a credible threat of unacceptable counteraction and/or belief that the cost of action outweighs the perceived benefits."[15] Some treat dissuasion, denial, and deterrence as separate concepts, but according to Adam Lowther, it can be useful to think of the three, along with threat, as operating across the spectrum of conflict with compellance intervening as a tool when deterrence fails.[16] Deterrence may be achieved through scalable levels of credible actions, beginning with dissuasion. Dissuasion does not threaten violence or punitive action but may take the form of influencing adversaries through diplomacy, international opinion, or national information campaigns to influence other nations. The next level of deterrence is denial, which seeks to deny the adversary its objective primarily through defensive measures. Deterrence by denial succeeds by decreasing the probability of success of an adversary. Examples of denial could include ballistic missile defenses (BMD), denial and deception operations, or increases in security and border protection. Deterrence by threat relies on the threat of use of a specific capability. Denial by threat can include overt threats that range from targeted strikes to an invasion by a conventional ground force. Threats can also incorporate punitive measures for noncompliance, including economic sanctions. If the adversary perceives its plans or intentions as being more costly than beneficial, then deterrence will succeed. If deterrence fails, punitive measures may be required using a compellance strategy to force a return to the previous status quo and avoid escalation.[17]

The two variables that matter most in deterrence are capability and will.[18] A persuasive threat of war backed up with a capable strategic arsenal may deter an aggressor, but the threat cannot appear to be bluffing or empty posturing. The perception of will and capability is also essential to credibility, and by extension, assurance, as the US continues to provide allies and partners security under the nuclear umbrella through extended deterrence agreements, both formal and implied, while discouraging them from developing their own nuclear weapons.

During the Cold War, deterrence focused on preventing nuclear war and nuclear proliferation and relied principally on a ready capability to retaliate against a Soviet surprise attack with a devastating response. The concept of central deterrence, supported by the assumptions and certainties of assured destruction, built stability into the relationship between the United States and the Soviet Union. The concept of extended deterrence provided security assurances for allies and partners under the US nuclear umbrella, thus discouraging them from developing their own nuclear weapons. During the Cold War, nuclear weapons were used to operationalize strategies of denial and punishment. Denial strategies, generally termed counterforce, focused on military targets, denying the adversary the ability to use its military forces, especially nuclear forces, in the event of a conflict. Punishment strategies, generally termed countervalue, focused on destroying the industrial capacity and urban centers of the adversary in order to impose unacceptable costs.

In the twenty-first century, an important aspect of deterrence planning will be to gain better insight into the strategic thinking of America's adversaries and understand their motivations to tailor deterrence strategies and develop credible messaging for more focused and effective deterrence strategies.[19] A more tailored approach to the three traditional elements of deterrence—threat, denial, and dissuasion—with an emphasis on designing deterrence strategies that hold at risk what an adversary values most, will have greater possibility for success.[20] The US must develop nuclear deterrence strategies that are tailored for each potential adversary, from our nearest peer competitor, Russia, to rogue states, potential nuclear proliferators, and non-state actors.[21]

The following table summarizes the quantitative and qualitative adjustments the United States must make to its nuclear triad and current deterrence strategies to confront a wide array of security challenges in the future strategic environment. It uses the Cold War as a reference point to highlight the changes in deterrence policy, nuclear force structure, and the addition of conventional weapons and theater missile defense that represent a new direction in US deterrence policy.

Twenty-first-century strategic threats

During the Cold War, the strategic arsenals of the United States and Soviet Union had a stabilizing effect on superpower relations and international stability by making any major conflict unacceptably risky. Although the risk of a nuclear exchange with Russia is now negligible, although growing because of Russian bellicosity, the American nuclear arsenal of the twenty-first century must maintain strategic stability with Russia and China, deter regional aggression, and prevent nuclear proliferation by assuring allies and partners that the nuclear umbrella remains credible.

In the future, the United States and its allies must also be prepared for conventional wars with nuclear-armed adversaries. Faced with a superior conventional force, a weaker adversary might threaten to use nuclear weapons to stop a war short of regime collapse and total defeat.[22] NATO successfully used the concept of coercive nuclear escalation during the Cold War when planning to

Table 1.1 Comparison of Cold War and twenty-first-century deterrence

	Cold War deterrent	*Twenty-first-century deterrent*
Threat	USSR	Multiple nuclear states, nuclear aspirants, nuclear proliferation, and nuclear terrorism.
US Nuclear Force Structure	Nuclear triad with over 10,000 warheads deployed on 2,000 launchers.	Nuclear triad with 1,550 strategic warheads deployed on 700 strategic delivery vehicles.
Deterrence Focus	Deterrence by counterforce and counter-value.	Deterrence by dissuasion, denial, threat, and compellance using nuclear and conventional deterrent forces.
Deterrence Policy	Central deterrence and extended deterrence.	Central deterrence, extended deterrence, and tailored deterrence.
Strategic Nuclear Force Mission	Maintain strategic stability with the USSR and assure US allies under US nuclear umbrella.	Maintain strategic stability with Russia and China, deter potential regional adversaries, and assure US allies and partners.
Stockpile Modernization/ Procurement	More generous budgets based on Cold War national security priorities.	Budgetary pressures will create significant risk if modernization and/or acquisition programs are delayed.
Allies and Partners	Assure under US nuclear umbrella.	Assure under US nuclear umbrella and US theater missile defenses.

Source: adapted from James M. S. Smith, "The New Strategic Framework, the New Strategic Triad, and the Strategic Military Services," in James J. Wirtz and Jeffrey A. Larsen, eds., *Nuclear Trans-formation: The New US Nuclear Doctrine* (New York: Palgrave Macmillan, 2005), 134; Hans M. Kristensen and Robert S. Norris, "Global Nuclear Weapons Inventories, 1945–2013," *Bulletin of the Atomic Scientists* (September 2, 2013): 76.

defend Europe from a superior Soviet conventional force. Any future adversary will likely consider the same strategy.[23]

There are now nine members of the nuclear club according to Hans Kristensen and Robert Norris.[24] Although the strategic nuclear forces of China, as well as India, North Korea, and Pakistan, are not equal to those of the United States and Russia, they complicate regional stability and increase nuclear force structure requirements in support of US extended deterrence agreements. All nine nuclear nations, with the exception of the United States and United Kingdom, have modernized or upgraded their nuclear arsenals.[25] China, France, India, Pakistan, Russia, and possibly Israel and North Korea, are likely to increase their nuclear weapons inventories, although none will reach parity with the United States and Russia for several decades unless both countries continue nuclear reductions as a result of additional bilateral agreements.

An overview of strategic nuclear forces and ballistic missile capabilities that may pose a threat to the United States and its allies, by tier according to number of warheads, delivery vehicles, and ballistic missile capabilities, along with a discussion of nuclear proliferation and nuclear terrorism threats, illustrates the complexity of the future US nuclear strategic environment.

Tier one: Russia and China

Russia remains the only US peer in nuclear deterrent capabilities. It operates a nuclear triad with approximately 1,800 operational warheads deployed on 326 ICBMs, 624 SLBMs on ten SSBNs, and 810 warheads for 60 strategic bombers; Russia is expected to comply with New START limits of 1,550 operationally deployed strategic nuclear weapons on 700 or fewer delivery vehicles.[26] Another 700 strategic warheads are in storage, along with 2,000 nonstrategic warheads (tactical nuclear weapons), probably maintained to confront threats from NATO and China. Russia is in the process of modernizing its nuclear triad, concentrating on its ICBM leg, and will replace its Soviet-era ballistic missiles with fewer, but improved, versions by 2023.[27] Russia successfully tested a new type of mobile ICBM in 2012 according to Russian press reports. The Russian SS-27 Mod 1, an ICBM designed to counter ballistic missile defense systems, is now deployed in silos in six regiments. In addition, Russian officials claim to be developing a new class of hypersonic vehicle to allow Russian strategic missiles to penetrate missile defense systems. The Russian press has indicated that acquisition of a new rail-mobile ICBM is under consideration.[28]

China currently has an operational nuclear dyad with roughly 250 warheads for 150 ICBMs, and a small inventory of air-delivered nuclear bombs.[29] China also has a nuclear weapons modernization program to achieve a nuclear triad, with SLBM production under way for three Jin-class SSBNs.[30] The number of Chinese ICBM nuclear warheads capable of reaching the continental US could expand to well over 100 by 2025, although the United States may not retaliate with ICBMs in response to a Chinese attack, since missiles launched from the central United States would overfly Russia to strike most potential targets in East Asia and the Middle East.[31] Estimates predict that China will add ten warheads annually to its nuclear inventory, depending on requirements for additional delivery vehicles.[32] China, the world leader in diverse ballistic missile progress, is developing advanced anti-access/area denial capabilities, including anti-ship ballistic missiles (ASBMs), that can threaten its neighbors and US forces deployed in the region.[33] China continues to field very large numbers of conventionally armed short-range ballistic missiles (SRBMs) opposite Taiwan.[34] Additionally, it is developing methods and weapons to counter US ballistic missile defenses. China is adding the DF-31A to the ICBM force. Future ICBMs may include MIRVs, depending on Indian plans to use MIRVs in the future.[35] According to press reports, China also recently tested a hypersonic glide vehicle that is intended to defeat ballistic missile defenses.[36]

Tier two: India and Pakistan

Indian and Pakistani warheads are in storage and not operationally deployed. The two countries primarily focus their deterrent on one another, although Indian long-range weapons are designed to deter China from aggression. Pakistan maintains 100–120 warheads for air and medium-range ballistic missile delivery systems and has considered producing a variety of miniature nuclear warheads that would allow it to arm anti-ship missiles as well as nuclear torpedoes, nuclear depth bombs, and nuclear naval mines.[37] Pakistan is also developing new delivery systems, to include a new nuclear-capable medium-range ballistic missile (MRBM), two new nuclear-capable SRBMs, and two new nuclear-capable cruise missiles.[38] Pakistan recently announced that it will develop its own SSBN to counter the Indian SSBN threat.[39]

India maintains 90–110 warheads for air, missile, and SSBN delivery systems, and is planning to increase its fissile material production capacity.[40] In 2012 India leased an SSBN from Russia for a period of ten years for use while it develops its own.[41] India conducted the first flight test of the Agni V ICBM in April 2012 and an even longer-range ICBM is reportedly in the design phase.[42] India is considering development of a MIRV capability for its ICBM, which combined with increased US missile defense capabilities in the Pacific region, could prompt China to do the same.[43] Pakistan continues to steadily expand its nuclear capabilities with the construction of two new plutonium production reactors and a new reprocessing facility.[44]

Tier three: North Korea and Iran

North Korea continues development of the TD-2 ICBM/Space Launch Vehicle (SLV), which could threaten the United States if developed as an ICBM. Launches in July 2006, April 2009, and April 2012 ended in failure, but a December 2012 launch successfully placed a satellite in orbit.[45] In 2012, North Korea unveiled the new but untested Hwasong-13 road-mobile ICBM which could also threaten the United States. In 2013, the Defense Intelligence Agency concluded, with moderate confidence, that North Korea may have developed a nuclear warhead small enough to be placed on a ballistic missile.[46]

While Iran has not yet developed its own nuclear weapons, it has an extensive missile development program that has received support from China, North Korea, and Russia.[47] The Iranian Shahab 3 MRBM, based on the North Korean No Dong missile, has been modified to extend its range and effectiveness, with the longest-range variant reportedly being able to reach targets at a distance of about 2,000 kilometers.[48] Iran has conducted multiple launches of the Sejjil, a solid-propellant MRBM with a claimed range of 2,000 km. In addition, it has conducted multiple launches of the Safir, a multi-stage SLV that serves as a test bed for long-range ballistic missile technologies.[49] Economic sanctions and international pressures have brought Iran to the negotiating table, but it continues its efforts to develop weapons-grade uranium and weapon delivery systems.[50]

Tier four: nuclear proliferation

There is clear evidence in diplomatic channels that US nuclear security assurances through extended deterrence agreements continue to be the single most important reason that 30 nations have foresworn nuclear weapons to date.[51] If there is serious doubt about the American nuclear umbrella, allies and partners may acquire their own nuclear arsenals. History has shown how difficult it is to compel a state to cancel a successful nuclear program once started. South Africa and Libya are success stories, but Syria continued its effort to join the nuclear club until its North Korean-built complex was destroyed by Israel.[52] Recent North Korean attempts to transport nuclear technology to other countries have been denied. Saudi Arabia was recently reported to be seeking a nuclear capability from Pakistan as a counter-balance to the threat from Iran, which could lead to a nuclear arms race in the region.[53] In addition to Saudi Arabia, Egypt, Turkey, and the United Arab Emirates could seek to acquire nuclear weapons since they believe that an Iran in possession of a nuclear deterrent might feel so safe from US or Israeli retaliation that it could act far more aggressively to dominate the Middle East and increase support to Hezbollah, Hamas, and other terrorist and insurgent groups.[54]

Agreements to support allies under the US BMD umbrella are also enhancing assurance credibility, and thus discouraging proliferation in the face of growing threats from short-range, medium-range, and intermediate-range ballistic missiles in regions where the United States maintains security relationships.[55]

Tier five: nuclear terrorism

Michael Levi in *On Nuclear Terrorism* discusses the practical difficulties terrorists face in acquiring and detonating a nuclear weapon.[56] Even if a terrorist group succeeded in buying or stealing sufficient fissile material, the undetected construction of a nuclear weapon or improvised nuclear device is probably beyond the technical capabilities of terrorists.[57] A terrorist group could steal a nuclear weapon, complete with ignition device, but would face difficulties in overcoming safing, arming, fusing, and firing (SAFF) procedures that could include required changes in altitude, acceleration, or other factors for detonation.[58] The easiest weapon to acquire is a non-nuclear device called a "dirty bomb," or radiological dispersal device (RDD), that theoretically would disperse radioactive material by combining it with conventional explosives.[59] Alternatively, a nuclear state could sell or provide a terrorist with a nuclear weapon and the means to detonate it, but there is little evidence to support this scenario, given the negative consequences that would result from an accidental detonation or retaliation against the state that supplied the weapon. Iran, Iraq, and Libya stopped sponsoring terrorist strikes against the United States after attacks were attributed to them.[60] A more likely scenario involves terrorist use of a nuclear weapon for purposes of blackmail or propaganda.[61] Finally, in a failed-state situation in Pakistan, a terrorist could gain access to nuclear weapons, but they would have to convince a group of trained military personnel to assist them to launch the weapon, which is unlikely due to fear of retribution or accident. Additional barriers to terrorist success in this scenario include: Pakistani SAFF features that prevent unauthorized use; separate nuclear storage facilities and delivery systems dispersed throughout Pakistan; and nuclear warheads that are stored unassembled, with cores separate from the weapons.[62]

Conclusion and recommendation

To cope effectively with the current and future multi-tiered threat environment, the United States must maintain an effective nuclear deterrent that is both capable and credible. This will require safe and effective nuclear weapons, new delivery systems, and tailored deterrence strategies that communicate the will of the United States to respond decisively to any aggression against itself or its allies and partners. The American nuclear triad provides the complementary mix of survivable, flexible, and responsive capabilities needed to support the range of options that may be required to confront multiple threats in the future—as long as modernization funding continues without interruption. Tailored deterrence strategies that hold at risk what each adversary values most will ensure that the nuclear triad can effectively provide strategic stability, discourage proliferation while assuring allies and partners, and deter regional aggression. The United States must make both qualitative and quantitative adjustments to its nuclear triad and current deterrence strategies in order to confront a wide array of security challenges in the future international nuclear strategic environment. Adjustments and changes in deterrence policy, nuclear force structure, and the addition of conventional weapons and theater missile defense represent a new direction in US deterrence policy required to support the twenty-first-century deterrence mission.

Recommendation: maintain the momentum for modernization

The past four years have been historic in terms of setting a new direction for the US nuclear enterprise. The United States has a new Nuclear Employment Strategy that confirms the fundamental role of the nuclear triad in national security. Proponents of the strategic nuclear force appear to have won for the time being. It is time to end the debate over triad legs and nuclear abolition in order to focus instead on funding to support nuclear modernization and procurement programs. Funding is finally available to modernize weapons or acquire new delivery and command and control systems but may soon become the biggest threat to the nuclear enterprise. The Monterey Institute of International Studies has received a lot of attention for their 2014 study "The Trillion Dollar Nuclear Triad."[63] The report estimates that the United States plans to spend approximately $1 trillion over the next 30 years to maintain the nuclear enterprise, procure replacement systems, and upgrade existing nuclear warheads.[64] Procurement of delivery systems and warheads will peak during a four- to six-year window after

2023 and may lead policymakers and lawmakers to delay funding for modernization and acquisition due to a mistaken perception of excessive cost during this period.[65] According to the study, the United States will actually spend only 3 percent of its defense budget on modernization efforts, which represents a very cost-effective effort in support of an alert nuclear deterrent that has suffered previously from delays and cancellations in upgrades due to budget constraints. Extending the "procurement holiday" for the nuclear enterprise could result in even higher future costs while undermining the credibility of the US nuclear deterrent. Under extreme budgetary pressures, policymakers might be forced to cancel one of the Air Force triad legs.[66] The United States must continue to fund NNSA modernization initiatives and acquisition of new delivery systems or be prepared to accept significant risk of technical failure as warheads and delivery systems age far beyond the dates they were designed to be effective.

The nuclear command, control, and communications (NC3) system is an outdated system of systems that includes numerous terrestrial, airborne, and space-based systems used to connect the president to US nuclear forces. Some systems were built in the 1960s and use vacuum tubes. Other systems use equipment and parts that are no longer manufactured and must be fabricated. In many cases, the systems are based on old analog technology. NC3 should be prioritized for replacement to support nuclear modernization and new delivery systems.

As subsequent chapters will establish, the need for a nuclear arsenal comprised of modern warheads and delivery platforms remains a persistent need. In spite of calls to the contrary, recent events in the Ukraine, South China Sea, and elsewhere suggests that the rapidly changing strategic environment often offer unpredictable change that, should the United States allow its nuclear arsenal to fall into further disrepair, will leave the nation at a distinct disadvantage against its adversaries. In short, the end of the Cold War has not sheparded in a world where nuclear weapons are anachronisms of the past.

Notes

1 George P. Shultz et al., "A World Free of Nuclear Weapons," *Wall Street Journal*, January 4, 2007, A15.
2 James Wood Forsyth Jr., B. Chance Salzman, and Gary Shaub Jr., "Remembrance of Things Past: The Enduring Value of Nuclear Weapons," *Strategic Studies Quarterly* (Spring 2010): 83. Forsyth, Salzman, and Shaub propose that the strategic arsenal could be reduced to a small number of counterforce and countervalue weapons totaling just over 300 for an effective minimum deterrent, and support preserving the three legs of the nuclear triad. They believe that China, India, Pakistan, North Korea, and, presumably, Iran understand that a small number of nuclear weapons would be all that is needed for deterrence to be effective if both civilian (countervalue) and military (counterforce) targets are part of the adversary's risk calculation. George Perkovich with the Carnegie Endowment for International Peace agrees that a total of 300 warheads would be a sufficient deterrent, but under two conditions: if China were to forgo further expansion of its nuclear arsenal, and if US friends and allies in Asia could be persuaded that reductions in US capabilities would not increase the threat of aggression from North Korea and China. George Perkovich, *Do Unto Others: Toward A*

Defensible Nuclear Doctrine (Washington, DC: Carnegie Endowment For International Peace, 2013), 66. Advocates of "Global Zero," the international movement for the eventual elimination of all nuclear weapons, support a nuclear arsenal of 900 total strategic nuclear weapons but propose eliminating non-strategic weapons along with the Minuteman intercontinental ballistic missile (ICBM) force over the next ten years. The notional force recommended would consist of 10 ballistic missile submarines armed with 720 warheads and 18 B-2 bombers armed with 180 gravity bombs. General James E. Cartwright (ret.) et al., *Global Zero US Nuclear Policy Commission Report: Modernizing US Nuclear Strategy, Force Structure and Posture* (New York: Global Zero, 2012), 6.

3 White House, Office of the Press Secretary, "Remarks by President Barack Obama," April 5, 2009, www.whitehouse.gov/the_press_office/Remarks-By-President-Barack-Obama-In-Prague-As-Delivered.

4 White House, Office of the Press Secretary, "Remarks by President Barack Obama."

5 US Department of Defense, *Nuclear Posture Review Report* (Washington, DC: US Department of Defense, 2010), 2.

6 US Department of Defense, *Nuclear Posture Review Report*, viii.

7 US Department of Defense, *Nuclear Posture Review Report*, 21.

8 US Department of State, "Treaty Between the United States of America and the Russian Federation on Measures for the Further Reduction and Limitation of Strategic Offensive Arms," April 8, 2010, 3. New START entered into force on February 5, 2011. It is expected to remain in effect until at least 2021. The United States agreed to limits by February 5, 2018, of 1,550 accountable strategic warheads, 700 deployed strategic delivery vehicles, and a combined limit of 800 deployed and non-deployed strategic launchers; to maintain the US nuclear triad of ICBMs, SLBMs, and nuclear-capable heavy bombers; and to modify ICBM payloads from existing multiple independently targetable reentry vehicle (MIRV) warheads to a single warhead each. The total number of deployed warheads may exceed the 1,550 limit by a few hundred because only one warhead per bomber is counted, regardless of how many it actually carries. Contributions by non-nuclear systems to US regional deterrence and reassurance goals are preserved by avoiding limitations on missile defenses and preserving options for using heavy bombers and long-range missile systems in conventional roles.

9 Jon Wolfstahl, Jeffrey Lewis, and Marc Quint, "The Trillion Dollar Nuclear Triad," James Martin Center for Nonproliferation Studies, 2014, 7. The White House agreed to request $88 billion for modernizing the nuclear complex and another $125 billion for sustaining and modernizing nuclear delivery systems.

10 Following the release of the 2010 NPR and ratification of the New START Treaty, the president directed DoD, the Department of State, Department of Energy (DOE), and the intelligence community to conduct an analysis of US nuclear deterrence requirements and policy in order to ensure US nuclear posture and plans are aligned to the twenty-first-century security environment. This study resulted in the Nuclear Weapons Employment Strategy of the United States.

11 Wolfstahl, Lewis, and Quint, "The Trillion Dollar Nuclear Triad," 7. The goal of a one-third reduction from New START levels to just over 1,000 nuclear weapons would be dependent on a verifiable bilateral agreement with the Russia Federation to cut its own stockpile to match US cuts.

12 US Department of Defense, *Report on Nuclear Employment Strategy of the United States Specified in Section 491 of 10 U.S.C.* (Washington, DC: Government Printing Office, 2013).

13 US Department of Defense, *Report on Nuclear Employment Strategy of the United States Specified in Section 491 of 10 U.S.C.*, 2.

14 US Department of Defense, *Report on Nuclear Employment Strategy of the United States Specified in Section 491 of 10 U.S.C.*, 5, 6. The stockpile "hedge" is required in case of technical failure of a critical system or additional requirements due to "unforeseen geopolitical changes."

15 US Department of Defense, Joint Publication 1–02, *Department of Defense Dictionary of Military and Associated Terms*, November 15, 2013, 73.

16 Adam Lowther, "Deterring Nonstate Actors," in *Thinking About Deterrence: Enduring Questions in a Time of Rising Powers, Rogue Regimes, and Terrorism*, ed. Adam Lowther (Maxwell AFB, AL: Air University Press, 2013), 199. These thoughts are based on concepts offered in: Edward Luttwak's *Strategy: The Logic of War and Peace*; Richard Kugler's "Dissuasion as a Strategic Concept"; Robert Pape's "Coercion and Military Strategy: Why Denial Works and Punishment Doesn't"; Daniel Byman and Matthew Waxman's "The Dynamics of Coercion: American Foreign Policy and the Limits of Military Might"; and Bernard Brodie's *Strategy in the Missile Age*.

17 Lowther, "Deterring Nonstate Actors," 199.

18 Thomas C. Schelling, *Arms and Influence* (New Haven, CT: Yale University Press, 1966), 35.

19 Jeffrey S. Lantis, "Strategic Culture and Tailored Deterrence: Bridging the Gap between Theory and Practice," *Contemporary Security Policy*, 30, 3 (December 2009), 467–485. For Elaine Bunn, a policymaker and strategist quoted by Lantis, this represents

> a shift from a one-size-fits-all notion of deterrence toward more adaptable approaches suitable for advanced military competitors, regional weapons of mass destruction states, as well as non-state terrorist networks.... [D]eterrence is about influencing the perceptions—and ultimately, the decisions and actions—of another party ... and may well differ in each circumstance or scenario.

20 Barry R. Schneider and Patrick D. Ellis, *Tailored Deterrence: Influencing States and Groups of Concern* (Maxwell AFB, AL: USAF Counterproliferation Center, 2012), 6.

21 Schneider and Ellis, *Tailored Deterrence: Influencing States and Groups of Concern*, 7.

22 As quoted in Perkovich, *Do Unto Others*, 17. The former Director of National Intelligence, Admiral Dennis Blair, addressed this possibility:

> The most likely circumstances of nuclear exchanges in these wars arise from American military superiority at the conventional level of war. With the United States on the way to victory, the governments of North Korea, China or Iran might threaten or actually use nuclear weapons to attempt to stop the war short of complete defeat.

General Sundarji from India reportedly remarked after the first Gulf War: "Never fight the Americans without a nuclear weapon." Countries ranging from North Korea to Iran to Pakistan have learned this lesson by observing US conventional operations since the first Gulf War.

23 Additional examples underscore the point:

- For example, if American and South Korean military forces were advancing on Pyongyang in response to conventional aggression, the North Koreans would likely use everything in their arsenal, to include nuclear weapons, to avoid defeat and regime change.
- If Iran decides to block the Strait of Hormuz, the Iranian regime could threaten nuclear retaliation if the United States decided to resolve the situation by force.
- Russia has already threatened to use nuclear weapons against Poland and NATO ballistic missile sites in the event of a large-scale conflict.

- If China decides to resolve its sovereignty claims in the South and East China Seas militarily, the US may be forced into a direct conflict with China in defense of US allies with counter-claims, to include Japan and the Philippines. China could defend against US "aggression" and avoid a naval confrontation by using space, cyber, or other assymetric weapons that could lead to escalation with the US.

Keir A. Lieber and Daryl G. Press, "The Nukes We Need: Preserving the American Deterrent," *Foreign Affairs*, 88, 6 (November/December 2009): 39–51; and Keir A. Lieber and Daryl G. Press, "The New Era of Nuclear Weapons, Deterrence, and Conflict," *Strategic Studies Quarterly*, 7, 1 (Spring 2013): 3–12. Lieber and Press argue that the idea of countries escalating conflict to avoid conventional defeat is well grounded in history.

24 Hans M. Kristensen and Robert S. Norris, "Global Nuclear Weapons Inventories, 1945–2013," *Bulletin of the Atomic Scientists* (September 2, 2013).
25 Kristensen and Norris, "Global Nuclear Weapons Inventories, 1945–2013," 75.
26 Hans M. Kristensen and Robert S. Norris, "Russian Nuclear Forces, 2013," *Bulletin of the Atomic Scientists* (May 2, 2013): 71.
27 Kristensen and Norris, "Russian Nuclear Forces, 2013," 71.
28 Kristensen and Norris, "Global Nuclear Weapons Inventories," 75.
29 Kristensen and Norris, "Global Nuclear Weapons Inventories," 75.
30 Hans M. Kristensen and Robert S. Norris, "Chinese Nuclear Forces, 2013," *Bulletin of the Atomic Scientists* (November 1, 2013): 80. China has two types of SLBMs for two types of SSBNs, but neither missile is operational.
31 National Air and Space Intelligence Center (NASIC), Public Affairs Office, *Ballistic and Cruise Missile Threat*, by NASIC, Defense Intelligence Agency, MSIC, and Office of Naval Intelligence (Wright-Patterson AFB, 2013), 3; and Evan B. Montgomery, *The Future of Americas Strategic Nuclear Deterrent* (Washington, DC: Center for Strategic and Budgetary Assessments, 2013), 21. This constraint raises the possibility that Moscow could mistake a nuclear strike against China as an attack on Russia and possibly discourage US policymakers from choosing to respond with ICBMs except under extreme circumstances.
32 Kristensen and Norris, "Chinese Nuclear Forces," 80.
33 NASIC, *Ballistic and Cruise Missile Threat*, 3.
34 NASIC, *Ballistic and Cruise Missile Threat*, 3.
35 NASIC, *Ballistic and Cruise Missile Threat*, 3.
36 Mike Hoffman, "Congress Reacts to Chinese Hypersonic Missile Test," *Defense Tech*, January 14, 2014.
37 Hans M. Kristensen and Robert S. Norris, "Pakistan's Nuclear Forces, 2011," *Bulletin of the Atomic Scientists* (July 1, 2011): 91.
38 NASIC, *Ballistic and Cruise Missile Threat*, 3.
39 Usman Ansari, "Pakistani Navy to Develop Nuclear-Powered Submarines: Reports," *Defense News*, February 11, 2012.
40 Kristensen and Norris, "Global Nuclear Weapons Inventories," 77.
41 Palash Ghosh, "India Joins Nuclear Submarine Community; Pakistan Alarmed," *International Business Times*, April 4, 2012.
42 NASIC, *Ballistic and Cruise Missile Threat*, 18.
43 Kristensen and Norris, "Global Nuclear Weapons Inventories," 77.
44 Kristensen and Norris, "Pakistan's Nuclear Forces," 91.
45 Kristensen and Norris, "Pakistan's Nuclear Forces," 91.
46 Thom Shanker, David Sanger, and Eric Schmitt, "Pentagon Finds Nuclear Strides by North Korea," *New York Times*, April 11, 2013. The Defense Intelligence Agency (DIA) assessment cautions that the nuclear weapon's "reliability will be low," due to technical issues with delivery systems and weapons design.
47 NASIC, *Ballistic and Cruise Missile Threat*, 14.

48 NASIC, *Ballistic and Cruise Missile Threat*, 14.
49 NASIC, *Ballistic and Cruise Missile Threat*, 14.
50 Ashish Sen and Douglas Ernst, "Iran Official: Sanctions 'Utterly Failed' to Stop Nuclear Program," *Washington Times*, December 4, 2013. In a recent interview, Iranian Foreign Minister Zarif boasted that "[w]hen sanctions started, Iran had less than 200 centrifuges. Today Iran has 19,000 centrifuges, so the net product of the sanctions has been about 18,800 centrifuges that have been added to Iran's stock of centrifuges."
51 US Department of State, International Security Advisory Board, *Report on Discouraging a Cascade of Nuclear Weapons States* (Washington, DC: Department of State, 2007), 22–23. "This umbrella is too important to sacrifice on the basis of an unproven ideal that nuclear disarmament in the US would lead to a more secure world.... A lessening of the US nuclear umbrella could very well trigger a cascade [of nuclear proliferation] in East Asia and the Middle East."
52 Lawrence Freedman, *Deterrence* (Cambridge, MA: Polity Press, 2004), 121.
53 Ali Ahmad, "The Saudi Proliferation Question," *Bulletin of the Atomic Scientists* (December 13, 2013): 1. In May 2012, Dennis Ross, a senior US diplomat and a former envoy to the Middle East, confirmed that in April 2009 King Abdullah explicitly told him, "If they [Iran] get nuclear weapons, we will get nuclear weapons."
54 Stephen Pifer et al., "U.S. and Extended Deterrence: Considerations and Challenges," *Brookings Arms Control Series*, Paper 3, The Brookings Institution, 2010, 523–557.
55 Stephen Pifer et al., "U.S. and Extended Deterrence: Considerations and Challenges," 523–557.
56 Michael A. Levi, *On Nuclear Terrorism* (Cambridge, MA: Harvard University Press, 2007), 27.
57 Levi, *On Nuclear Terrorism*, 28, 30, 35. Studies suggest that the price of a nuclear weapon, several million dollars, would be too high for today's terrorist groups. Additionally, the financial transaction required for payment has a high probability for discovery.
58 Levi, *On Nuclear Terrorism*, 125.
59 Nuclear Regulatory Commission, "Fact Sheet on Dirty Bombs," December 27, 2012, 1, www.nrc.gov/reading-rm/doc-collections/fact-sheets/fs-dirty-bombs.html. The primary impact of a radiological dispersal devise would be mass panic and terror since there would not be enough concentrated radiation in the affected area to cause severe illness or death.
60 Freedman, *Deterrence*, 121.
61 Adam Lowther, *Challenging Nuclear Abolition* (Maxwell AFB, AL: Air Force Research Institute, 2009), 18.
62 Kristensen and Norris, "Pakistan's Nuclear Forces," 94.
63 Wolfstahl, Lewis, and Quint, "The Trillion Dollar Nuclear Triad," 4.
64 Wolfstahl, Lewis, and Quint, "The Trillion Dollar Nuclear Triad," 4.
65 Wolfstahl, Lewis, and Quint, "The Trillion Dollar Nuclear Triad," 4.
66 Wolfstahl, Lewis, and Quint, "The Trillion Dollar Nuclear Triad," 4.

Bibliography

Ahmad, Ali, "The Saudi Proliferation Question," *Bulletin of the Atomic Scientists* (December 13, 2013).

Ansari, Usman, "Pakistani Navy to Develop Nuclear-Powered Submarines: Reports," *Defense News*, February 11, 2012.

Cartwright (ret.), General James E. et al. *Global Zero US Nuclear Policy Commission Report: Modernizing US Nuclear Strategy, Force Structure and Posture* (New York: Global Zero, 2012).

Forsyth Jr., James Wood, Salzman, B. Chance, and Shaub Jr., Gary, "Remembrance of Things Past: The Enduring Value of Nuclear Weapons," *Strategic Studies Quarterly* (Spring 2010).

Freedman, Lawrence, *Deterrence* (Cambridge, MA: Polity Press, 2004).

Ghosh, Palash, "India Joins Nuclear Submarine Community; Pakistan Alarmed," *International Business Times*, April 4, 2012.

Hoffman, Mike, "Congress Reacts to Chinese Hypersonic Missile Test," *Defense Tech*, January 14, 2014.

Kristensen, Hans M., and Norris, Robert S., "Chinese Nuclear Forces, 2013," *Bulletin of the Atomic Scientists* (November 1, 2013).

Kristensen, Hans M., and Norris, Robert S., "Global Nuclear Weapons Inventories, 1945–2013," *Bulletin of the Atomic Scientists* (September 2, 2013).

Kristensen, Hans M., and Norris, Robert S., "Pakistan's Nuclear Forces, 2011," *Bulletin of the Atomic Scientists* (July 1, 2011).

Kristensen, Hans M., and Norris, Robert S., "Russian Nuclear Forces, 2013," *Bulletin of the Atomic Scientists* (May 2, 2013).

Lantis, Jeffrey S., "Strategic Culture and Tailored Deterrence: Bridging the Gap between Theory and Practice," *Contemporary Security Policy*, 30, 3 (December 2009).

Levi, Michael A., *On Nuclear Terrorism* (Cambridge, MA: Harvard University Press, 2007).

Lieber, Keir A., and Press, Daryl G., "The New Era of Nuclear Weapons, Deterrence, and Conflict," *Strategic Studies Quarterly*, 7, 1 (Spring 2013).

Lieber, Keir A., and Press, Daryl G., "The Nukes We Need: Preserving the American Deterrent," *Foreign Affairs*, 88, 6 (November/December 2009).

Lowther, Adam, *Challenging Nuclear Abolition* (Maxwell AFB, AL: Air Force Research Institute, 2009).

Lowther, Adam, "Deterring Nonstate Actors," in *Thinking About Deterrence: Enduring Questions in a Time of Rising Powers, Rogue Regimes, and Terrorism*, ed. Adam Lowther (Maxwell AFB, AL: Air University Press, 2013).

Montgomery, Evan B., *The Future of Americas Strategic Nuclear Deterrent* (Washington, DC: Center for Strategic and Budgetary Assessments, 2013).

National Air and Space Intelligence Center (NASIC), Public Affairs Office, *Ballistic and Cruise Missile Threat*, by NASIC, Defense Intelligence Agency, MSIC, and Office of Naval Intelligence (Wright-Patterson AFB, 2013).

Nuclear Regulatory Commission, "Fact Sheet on Dirty Bombs," December 27, 2012, www.nrc.gov/reading-rm/doc-collections/fact-sheets/fs-dirty-bombs.html.

Perkovich, George, *Do Unto Others: Toward A Defensible Nuclear Doctrine* (Washington, DC: Carnegie Endowment For International Peace, 2013).

Pifer, Stephen et al., "U.S. and Extended Deterrence: Considerations and Challenges," *Brookings Arms Control Series*, Paper 3, The Brookings Institution, 2010.

Schelling, Thomas C., *Arms and Influence* (New Haven, CT: Yale University Press, 1966).

Schneider, Barry R., and Ellis, Patrick D., *Tailored Deterrence: Influencing States and Groups of Concern* (Maxwell AFB, AL: USAF Counterproliferation Center, 2012).

Sen, Ashish, and Ernst, Douglas, "Iran Official: Sanctions 'Utterly Failed' to Stop Nuclear Program," *Washington Times*, December 4, 2013.

Shanker, Thom, Sanger, David, and Schmitt, Eric, "Pentagon Finds Nuclear Strides by North Korea," *New York Times*, April 11, 2013.

Shultz, George P. et al., "A World Free of Nuclear Weapons, *Wall Street Journal*, January 4, 2007, A15.

US Department of Defense, Joint Publication 1–02, *Department of Defense Dictionary of Military and Associated Terms*, November 15, 2013.

US Department of Defense, *Nuclear Posture Review Report* (Washington, DC: US Department of Defense, 2010).

US Department of Defense, *Report on Nuclear Employment Strategy of the United States Specified in Section 491 of 10 U.S.C.* (Washington, DC: Government Printing Office, 2013).

US Department of State, International Security Advisory Board, *Report on Discouraging a Cascade of Nuclear Weapons States* (Washington, DC: Department of State, 2007).

US Department of State, "Treaty Between the United States of America and the Russian Federation on Measures for the Further Reduction and Limitation of Strategic Offensive Arms," April 8, 2010.

White House, Office of the Press Secretary, "Remarks by President Barack Obama," April 5, 2009, www.whitehouse.gov/the_press_office/Remarks-By-President-Barack-Obama-In-Prague-As-Delivered.

Wolfstahl, Jon, Lewis, Jeffrey, and Quint, Marc, "The Trillion Dollar Nuclear Triad," James Martin Center for Nonproliferation Studies, 2014.

Incomplete source citations

These thoughts are based on concepts offered in: Edward Luttwak's *Strategy: The Logic of War and Peace*; Richard Kugler's "Dissuasion as a Strategic Concept"; Robert Pape's "Coercion and Military Strategy: Why Denial Works and Punishment Doesn't"; Daniel Byman and Matthew Waxman's "The Dynamics of Coercion: American Foreign Policy and the Limits of Military Might"; and Bernard Brodie's *Strategy in the Missile Age*.

2 Is there future utility in nuclear weapons?

Nuclear weapons save lives

Robert A. Hoskins

Introduction

The political utility of nuclear weapons has come under increased scrutiny since the end of the Cold War. With the Soviet Union and the United States no longer at an alert standoff, some believe nuclear weapons no longer serve any purpose relative to national security. This is not the case. Those who believe in nuclear disarmament are discounting the historical reality that the world has been safe from great power war since nuclear weapons have become part of the military arsenal. Not only have nuclear weapons deterred nuclear war, but they have, and will continue to deter large-scale conventional war in a dangerous world. This chapter will examine the political utility of nuclear weapons in three sections.

In the first section, there is an examination of current arguments in support of nuclear disarmament. There are myriad voices, authors, and think tanks pushing an agenda of a global nuclear "zero." Consideration will be given to those who believe nuclear weapons no longer serve a purpose, and to those who believe the United States should lead the world into nuclear total disarmament. Their prevailing arguments will be outlined and analyzed. There are those who argue that nuclear weapons only deter nuclear war. There are also those who believe nuclear weapons are not a cost-effective portion of the military arsenal, and that American treasure should be spent elsewhere. Third, some believe nuclear weapons create instability in the world, and that nuclear proliferation, particularly among rogue states and/or violent non-state actors is the greatest threat to US national security in the current day. Effectively, this is the idea that nuclear weapons make the world a more dangerous place. Last, and potentially most important, there are those who support the idea that the American people could never stomach the use of nuclear weapons. It is an argument with some credibility. Why should the US have nuclear weapons if it will not use them? With the arguments in support of nuclear disarmament established, an examination in response to each will be offered.

The second section of this chapter provides counter arguments to the "views of others" outlined in the first section. First, an analysis of history will examine the effects of nuclear weapons. By reviewing the history of war in the twentieth century, an assessment of the utility of nuclear weapons is offered. Who has nuclear weapons? What have been the effects? Next, there is a review of the cost of continuing to maintain a nuclear arsenal in relation to large-scale conventional conflict and other American spending. This comparison sheds light on the country's perceived priorities. Third, the idea that nuclear weapons make the world unstable and/or unsafe, leading to greater potential for conflict, is reviewed. There is a review of historical case studies, potential for future threats, and the insinuated effect of nuclear weapons. Last, the argument that the United States will never use nuclear weapons is important. Should they remain an option for the national command authority? Following this section in support of the political utility of nuclear weapons, this chapter concludes with policy recommendations, such as whether to maintain a nuclear arsenal, the ramifications of doing so, and policy changes needed going forward.

Policy recommendations start with the idea that the United States must maintain a safe, secure, and effective nuclear deterrent in the future. This chapter will recommend that the United States modernize and dedicate national treasure to the nuclear arsenal as perhaps the most cost-effective deterrent to large-scale war and nuclear war. Furthermore, it is suggested that the United States nuclear arsenal is responsible for controlling some, although not all, instability in the world. There is also a need to assure allies and attribute nuclear weapons to their source in the event of their use, which are critical components to maintaining stability and security burden-sharing. Last, the recommendation is made that the debate over the nuclear arsenal needs to be opened up to a greater audience by the Department of Defense and the President. A deliberate effort to facilitate a national discussion is necessary. The United States's nuclear posture and its policy statements need to send a message to the world that clearly establishes credibility, is supported by the American people, and clearly communicates the American position to the rest of the world.

Views of others

Some believe nuclear weapons exist for the sole purpose of deterring nuclear war. They believe nuclear weapons possess no political utility beyond that function, and as a logical follow-on, that in a world without nuclear weapons, no one would require them for security. They argue that the existence of nuclear weapons is destabilizing the world over, and makes international security more challenging and small-scale military adventurism more likely. Robert Jervis described the situation as follows: "To the extent that the military balance is stable at the level of all-out nuclear war, it will become less stable at lower levels of violence."[1] The idea is that states will not be deterred from lower levels of war, and will operate with relative impunity short of large-scale conflict. Some point to a nuclear India and Pakistan as an example of instability. In 1997, South Asia observer Neil Joeck argued that

> India and Pakistan's nuclear capabilities have not created strategic stability [and] do not reduce or eliminate factors that contributed to past conflicts,...

Far from creating stability, these basic nuclear capabilities have led to an incomplete sense of where security lies. Nuclear weapons may make decision-makers in New Delhi and Islamabad more cautious, but sources of conflict immune to the nuclear threat remain. Limited nuclear capabilities increase the potential costs of conflict, but do little to reduce the risk of it breaking out.[2]

Effectively, the argument is that nuclear weapons do not stop war. This is correct. A historical review from the inception of the Atomic Age until the present is replete with examples of states, both nuclear and non-nuclear, engaging in conflict. Finally, there are those who believe the existence of nuclear weapons not only work against stability, but that nuclear weapons make nuclear war more likely. George Perkovich of the Carnegie Endowment writes, "If major powers of the twenty-first century are to avoid the destructiveness of the twentieth century, leaders will have to concentrate actively and assiduously on removing the temptation to initiate use of nuclear weapons."[3] The temptation Perkovich refers to is the existence of nuclear weapons.

In a time of economic austerity, some have voiced concern that nuclear weapons are not a cost-effective component of national defense. It is logical that military professionals, if left to their own devices and unfettered by budgets, want every advantage available. But with a current budget crisis and a spiraling national debt, is not the national treasure better spent elsewhere? Maintenance of the nuclear weapons complex costs approximately $25–30 billion per year, and it is estimated to cost $179 billion between 2010 and 2018, and then balloon to $500 billion over the next 20 years.[4] While these numbers are debatable, they are in the neighborhood of the general consensus. Clearly, nuclear weapons are not trivial public expenditures. Maintenance costs can be dubious and ambiguous when one considers there is more involved in the cost of these weapons. Delivery systems, including high-ticket items like nuclear submarines, installation infrastructure, personnel costs, and other requirements drive the price tag for the nuclear enterprise. Considering the fact that the United States has not launched a nuclear attack since 1945, it is not surprising some argue that the cost of nuclear weapons is prohibitive. The Nuclear Weapons Inheritance Project suggests: "The costs of nuclear weapons programs is enormous and for every dollar invested in advanced weapon systems a dollar less is invested in health, education, social welfare and development."[5] With a fixed budget, it is clear that funding nuclear weapons is a trade-off with other spending. It further points to more far-reaching financial trade-offs, and argues that

the price of global elimination of starvation, provision of health care, provision of shelter and clean water, elimination of illiteracy, provision of sustainable energy, debt relief for developing countries, clearance of landmines and more has been estimated to be about $260 billion annually for 10 years.[6]

Clearly, the maintenance of the nuclear arsenal is a trade-off. In fact, the nuclear arsenal costs more annually than the individual gross domestic product of the world's bottom 90 nations.[7] Do Americans need a nuclear arsenal that costs billions of dollars per year, and is projected to cost even more going forward, when we could redirect that money to health care, worldwide development, clean technology, and other priorities? Perhaps the taxpayers could do better.

In his May 6, 2009, Prague speech, the President of the United States said that "we must ensure that terrorists never acquire a nuclear weapon. This is the most immediate and extreme threat to global security." He further stated, "The existence of thousands of nuclear weapons is the most dangerous legacy of the Cold War."[8] The emergence of international, violent non-state actors has presented a challenge. Rogue states are also a concern. Iran is pursuing a nuclear weapons capability, and has established relationships with terrorist groups like Hezbollah. The rationality of North Korea's leadership, and their compulsion to avoid becoming a responsible member of the international community is problematic. In his book, *On Nuclear Terrorism*, Michael Levi asserts, "Theft is not the only way to acquire nuclear weapons or materials—states or their senior officials might deliberately transfer nuclear weapons or materials to terrorist groups."[9] Failed states or potentially failed states could also be a target of opportunity for terrorists to acquire nuclear weapons. A nuclear-armed Pakistan, faced with internal instability, could lapse in its nuclear security to the point that terrorists from the region could acquire weapons or material. These are considerable concerns, and warrant examination. President Obama was not exaggerating when he claimed, "One nuclear weapon exploded in one city—be it New York or Moscow, Islamabad or Mumbai, Tokyo or Tel Aviv, Paris or Prague—could kill hundreds of thousands of people."[10] Assuming a yield consistent with modern-day nuclear weapons, a nuclear detonation in a major city would be the most horrific act of instantaneous violence in the history of mankind.

Last, there is the argument that the United States will never again use nuclear weapons. They no longer provide a credible deterrent. In the 2010 *Nuclear Posture Review Report*, the United States announced "negative security assurance by declaring that the United States will not use or threaten to use nuclear weapons against non-nuclear weapons states that are party to the NPT and in compliance with their nuclear non-proliferation obligations." The report went further in stating that

> any state eligible for the assurance that uses chemical or biological weapons against the United States or its allies and partners would face the prospect of a devastating conventional military response—and that any individuals responsible for the attack, whether national leaders or military commanders, would be held fully accountable.

And finally, the report allowed the possibility that for

> states that possess nuclear weapons and states not in compliance with their nuclear non-proliferation obligations—there remains a narrow range of contingencies in which US nuclear weapons may still play a role in deterring a

conventional or CBW (chemical, biological weapons) attack against the United States or its allies and partners.[11]

While the preceding was a litany of "what if" statements, it indicates a prevailing reluctance to employ nuclear weapons. In fact, it strategically communicates a very narrow set of circumstances in which the United States would consider using nuclear weapons. For an adversary, it is a roadmap to American redlines. Considering the range of military options available to worldwide state and non-state actors alike, this posture effectively confirms that short of nuclear attack, use of CBW, or large-scale conventional attack by a near peer, the United States's nuclear arsenal will stand down. Considering the likelihood of such an attack, why maintain such an arsenal? The United States's own nuclear posture and policy nearly dictates that the country's leadership is moving further and further away from the nuclear option under any circumstances. The stated policy of the United States government implies that nuclear weapons may eventually be a sunset capability, although current policy guidance calls for modernization of the nuclear arsenal and infrastructure. As such, the credibility of the nuclear deterrent has been overtaken by the pursuit of conventional superiority. The United States will not use nuclear weapons except under very restrictive circumstances, so why maintain the capability?

Counter arguments

The views of nuclear skepticism outlined above, including statements made by the American officials, are powerful and resonate in leadership circles around the world. There is an alternative view that demands attention in a dangerous world. Nuclear weapons have saved lives and will continue to save lives in the future. They will continue to exert political utility if managed, maintained, postured, and communicated correctly, and they are vital to US national security.

Prior to discussing views in support of the nuclear arsenal, it is important to establish historical background salient to the issue. Who has nuclear weapons, when did they acquire them, and what do they have? Table 2.1 gives a short outline of worldwide nuclear powers.

This background is important as we review the opposing viewpoints, as it can be instructive as to state behavior, stability, and security concerns. The first argument to address is that nuclear weapons only deter nuclear war. Many opponents of nuclear weapons misperceive the utility of them. Nuclear weapons do not exist to stop all war, just as a shotgun is not meant for killing spiders in one's home. Historically, nuclear weapons have accomplished two things: deterred nuclear war and de-escalated or averted great-power war. This is historically supported by a review of twentieth-century deaths resulting from war. Prior to the culmination of World War II, the world went through the first part of the twentieth century with tens of millions of war casualties as a result of great-power war. Following World War II, there has been a drastic decline in worldwide war-related deaths, and only one factor has changed, the advent of nuclear

Table 2.1 Worldwide nuclear powers

Country	First detonation	Warheads
United States	1945	1,654 deployed*
USSR	1949	1,480 deployed*
UK	1952	225
France	1960	300
China	1964	240
India	1998	Approx. 100
Pakistan	1998	Approx. 100
North Korea	2006	Approx. 5
Israel	1979?**	75–200

Notes
* The United States and Russia maintain weapons in "deployed" status, as well as reserves.
** Israel is known to have nuclear weapons, but does not have a confirmed test. There is conjecture that Israel participated in a joint test with South Africa in 1979.[12]

weapons. During World War I and World War II, exclusively, war dead numbered over 25 million military members. Adding in civilians to the total for World War II brings the total count to nearly 70 million.[13] In the period following World War II to the present, war dead worldwide in all manner of conflict has been about 3.7 million.[14] This presents a significant contrast. There remain great powers with great militaries and opposing national interests. There remains evil in the world. The change has been that the cost of war has risen among the great powers to the point it is potentially unwinnable, due to the nuclear option. Evidence from the Korean War[15] and the Vietnam War[16] support the view that the potential for nuclear power escalation played a role in the decision makers' calculus on both sides. In both cases, great-power war was averted, and while major powers supported opposing sides of these conflicts, they avoided large-scale war with one another. Further, and more recently, India and Pakistan's relationship has proven that nuclear weapons are de-escalatory. As is often the case in the nuclear debate, parties can view the same circumstances through an entirely different lens. J. N. Dixit, the national security advisor to former Indian Prime Minister Manmohan Singh wrote, "A certain parity in nuclear weapons and missile capabilities will put in place structured and mutual deterrents. These could persuade the Governments of India and Pakistan to discuss bilateral disputes in a more rational manner." Further, India's Army Chief, K. Sundarji, predicted that "the only salvation is for both countries to follow policies of cooperation and not confrontation.... A mutual minimum nuclear deterrent will act as a stabilizing factor."[17] Clearly, nuclear weapons possess utility beyond deterring nuclear war.

Having discussed the utility of nuclear weapons, are they cost prohibitive? The answer to this question is a matter of perspective. Considering the fact that $25–30 billion a year is about 5 percent of US annual military defense spending, nuclear weapons are a bargain. However, considering potential trade-off spending, this is a question of priorities and perspective. The American people supported World War

II without an internal revolution to the tune of what would be trillions of modern-day dollars. It is arguable whether or not the Axis powers presented an existential threat to the United States, but the threat to the country and its allies was considerable. Russian and Chinese nuclear weapons present an existential threat to the United States on a daily basis. They each have enough weapons to destroy the American way of life within minutes. There is no arguing that $25–30 billion is a significant national expenditure, but it is far less expensive than great-power war or large-scale conventional war. The Iraqi and Afghanistan wars have cost the country in excess of two trillion dollars, and counting, and they do not present an existential threat. The question for the American people: Are they comfortable abandoning nuclear weapons for budgetary savings in the face of other countries with the capability of destroying the United States? It is more a question of priorities, not cost. The United States can afford its nuclear weapons program, which serves to keep the country safe from capable and malicious adversaries.

The potential for nuclear terrorism sponsored by rogue states or as a result of a nuclear security failure also requires attention. While President Obama declared a nuclear weapon in the hands of terrorists the greatest threat facing America—which is not the case—it is not an existential threat. Chinese and Russian nuclear weapons are the greatest existential threats to American security.[18] The United States has the ability to cooperate with other nuclear states, like Pakistan, to enhance their nuclear security. Security enhancements on the weapons themselves, as well as process improvements and communication protocols between nuclear adversaries work to that end. Additionally, nuclear forensics leading to attribution of sources for nuclear weapons or fissile materials are critical and attainable. Levi describes nuclear forensics as the "science and art linking nuclear materials to their sources."[19] The United States has the ability and can communicate the ability to attribute the origin of nuclear weapons or material to their source. By partnering with states on nuclear security, and by clearly communicating the ability to attribute the origin of nuclear materials, the United States can hold state actors at risk and deter proliferation to violent non-state actors. It is important to reiterate that violent non-state actors are not deterred by America's nuclear arsenal, and should they obtain and detonate a device, it would be horrific. States are deterred by the nuclear arsenal, and states present existential threats, terrorists do not. The President's comment that the "existence of thousands of nuclear weapons is the most dangerous legacy of the Cold War" is also inaccurate. Thousands of secure nuclear weapons are safer than one unsecure weapon.[20] The legacy of the Cold War is that nuclear deterrence works. Further, a safe, secure arsenal, large enough to deter potential adversaries and assure allies, mitigates proliferation. As allied countries hear presidential rhetoric about a desire to shrink the nuclear arsenal, or restrictive interpretations of US nuclear employment policy, they may incorrectly fear depreciation in US commitment to extended nuclear deterrence.

Is there any veracity to the contention that the United States will almost never use nuclear weapons? The issue of credibility must be addressed. Once capability is established, credibility becomes more of a discussion about intentions. The critical component to the credibility of the nuclear arsenal is not what the United States will actually do when challenged, but rather, what other states believe the United States will do. Credibility is a perception issue. Karl Heinz-Kamp and David S. Yost in "NATO and 21st Century Deterrence" describe credibility as "the interplay of capability and resolve."[21] The US maintains nuclear credibility in the eyes of potential adversaries. The United States is the only nation that has used nuclear weapons in war. Adversaries remember that. Further, the idea that America has not used nuclear weapons since 1945 is incorrect. Nuclear weapons have been used every day since 1945 to provide a strategic deterrent. Evidence of this is that the United States has not had to kinetically engage an existential threat since World War II. In fact, since America's reaction to the attacks of September 11, 2001, the prevailing perception is that the US is prepared to hold adversaries at great risk when presented with a threat to the homeland. If one expands this threat to the nuclear realm, there can be no doubt that the US maintains credibility. Consider Israel, a nation surrounded by adversarial countries. Israel has thwarted conventional attacks from neighboring countries, but has never faced a battle for survival. If those countries did not believe Israel would resort to nuclear weapons in the face of a threat to national survival, Israel may have faced greater aggression. While Israel has never officially detonated a nuclear device, it maintains a credible deterrent due to the perception of nuclear capability. Deterrence is in the eye of the beholder, and American's nuclear arsenal remains credible, and has history to back that up.

Policy recommendations

Presently, the United States continues to maintain a nuclear arsenal. The future of that arsenal, and current debates surrounding it, demand a policy review. The US President's principal policy statement must be an overt commitment to maintaining a nuclear arsenal large enough to provide deterrence of potential adversaries against nuclear attack or blackmail of the United States and its allies, to provide reassurance of nuclear security for the US nuclear weapons complex and arsenal, and to support nuclear employment flexibility and resilience under all conditions of threat or, if necessary, nuclear use. The exact figure is unknown, but at some possible future point, a reduced American arsenal could force partner states to abandon the extended deterrence umbrella and proceed with their own nuclear programs. Further, it is critical that in official United States communications, government officials exhibit a commitment to the nuclear deterrent for the future. The message should be clear: The United States is posturing and resourcing itself based on the capability of other states and the potential threat, not in perceived intent. Intent can change. It would equate to military malpractice to recommend mitigation of the nuclear deterrent in the face of existential threats to the nation.

In addition to committing to maintenance of the nuclear arsenal, the United States should modernize its nuclear capability. The country cannot allow its sole deterrent against an existential threat to rust into retirement. Russia, China, and

France have modernized weapons, and the United States has remained politically constrained from pursuing new capability. To illustrate this point, two current debates are under way regarding a new bomber program and a redesign of the tail kit for the bombs they could deliver. The B-52 fleet is nearly 60 years old, and the B-61 bomb it delivers is over 50 years old. Nuclear weapons remain the most cost-effective military capability the nation possesses. Not only would modernization and increased investment assure capability and enhance credibility, but it would send a message to potential adversaries that the US will continue to hold them at risk should their intent turn malicious. Nuclear weapons are fundamentally a self-defense capability. There is no historical example for when a weaker US made the world a safer place. If the country is committed to maintaining the American way of life, it will demand commitment to nuclear weapons.

The United States has a decision to make between the safety derived from leadership, or the hope of a more docile world. Absent leadership, proliferation may occur horizontally. Current non-nuclear states may pursue nuclear capability in response to US conventional superiority or in response to decreases in the US arsenal. The US must exert international security leadership and ensure cooperation and discipline among nuclear nations. To a degree, proliferation among allies is within US control. Maintaining an effective, safe, and secure arsenal assures allies. Ensuring states understand the existence of nuclear forensics and attribution will also dictate behavior. America cannot be a shrinking power, communicating weakness and false promises of nuclear abstinence that may embolden nuclear-aspiring rogue states.

There is no doubt the nuclear debate will continue. However, the debate requires expansion. This discussion is too important to be left among think tanks and policy professionals, as it potentially affects the lives of all Americans. The US has public debates regarding health care, steroid use in baseball, and American Idol, but not about nuclear weapons? Because it is not rational to pursue unilateral disarmament in the face of a world with nuclear weapons, this has to be a discussion laid before the American people by the Department of Defense and the president. During the Cold War, Americans understood deterrence. There were public service announcements regarding actions to take during a possible attack, nuclear drills for children at school, and fall-out shelters in government buildings for citizens. This is not a recommendation to return to nuclear paranoia, but rather, to remind the American public why it is critical to maintain a nuclear arsenal for which they dedicate $25 billion a year. The threat and the solution must be supported by the American people, and they have been left out of the discussion. Americans have seen news stories regarding nuclear weapon buffoonery. They have seen wars in recent years that, while costly in terms of life and treasure, did not impact their daily ability to attend ballet practice and space camp. They need to be reacquainted with the sobering reality that there is a threat requiring their attention, and a solution requiring their support.

Notes

1 Robert Jervis, *The Illogic of American Nuclear Strategy* (Ithaca, NY: Cornell University Press, 1984), 31.
2 Neil Joeck, "Maintaining Nuclear Stability in South Asia," *Adelphi Paper 312*, 1997, 12.
3 George Perkovich, *Do Unto Others* (Washington, DC: Carnegie Endowment for International Peace, 2013), 78.
4 Nuclear Threat Initiative, "Nuclear Weapons Budget Overview," www.nti.org/analysis/articles/us-nuclear-weapons-budget-overview/.
5 International Physicians for the Prevention of Nuclear War, www.ippnw-students.org/NWIP/pdfs/costs.pdf.
6 International Physicians for the Prevention of Nuclear War.
7 CIA World Fact Book 2003–2010, "List of Countries by GDP," http://en.wikipedia.org/wiki/List_of_countries_by_GDP_(nominal).
8 Barack H. Obama, address, Prague, Czechoslovakia, April 5, 2009.
9 Michael Levi, *On Nuclear Terrorism* (Cambridge, MA: Harvard University Press, 2007), 127.
10 Obama, address, Prague, Czechoslovakia.
11 Department of Defense, *Nuclear Posture Review Report* (Washington, DC: Government Printing Office, 2010), vii.
12 Arms Control Association, "Nuclear Weapons, Who Has What at a Glance," www.armscontrol.org.
13 Antony Beevor, "The Second World War," *The Economist*, June 9, 2012, www.economist.com/node/21556542.
14 The Correlates of War, "Data Sets of Inter-state Wars," www.correlatesofwar.org.
15 General Omar N. Bradley, Chairman, Joint Chiefs of Staff, to Louis A. Johnson, Secretary of Defense, memorandum, July 10, 1950, 2.
16 George W. Ball, Undersecretary of State, to Chairman Joint Chiefs of Staff and Secretary of State, 1964, 23, 33.
17 Michael Krepon, "The Stability-Instability Paradox," *Prospects for Peace in South Asia* (Palo Alto, CA: Stanford University Press, 2004), 4.
18 Adam Lowther, *Challenging Nuclear Abolition* (Maxwell AFB, AL: Air Force Research Institute Papers, 2009), 19.
19 Levi, *On Nuclear Terrorism*, 127.
20 Lowther, *Challenging Nuclear Abolition*, 16.
21 Karl Heinz-Kamp and David S. Yost, eds., "NATO and 21st Century Deterrence," NATO Defense College Paper 8, 2009, 161.

Bibliography

Arms Control Association, "Nuclear Weapons, Who Has What at a Glance," www.armscontrol.org.
Ball, George W., Undersecretary of State, to Chairman Joint Chiefs of Staff and Secretary of State, 1964.
Beevor, Antony, "The Second World War," *The Economist*, June 9, 2012, www.economist.com/node/21556542.
Bradley, General Omar N., Chairman, Joint Chiefs of Staff, to Louis A. Johnson, Secretary of Defense, memorandum, July 10, 1950.
CIA World Fact Book 2003–2010, "List of Countries by GDP," http://en.wikipedia.org/wiki/List_of_countries_by_GDP_(nominal).
The Correlates of War, "Data Sets of Inter-state Wars," www.correlatesofwar.org.

Department of Defense, *Nuclear Posture Review Report* (Washington, DC: Government Printing Office, 2010), vii.

Heinz-Kamp, Karl, and Yost, David S., eds., "NATO and 21st Century Deterrence," NATO Defense College Paper 8, 2009.

International Physicians for the Prevention of Nuclear War, www.ippnw-students.org/NWIP/pdfs/costs.pdf.

Jervis, Robert, *The Illogic of American Nuclear Strategy* (Ithaca, NY: Cornell University Press, 1984).

Joeck, Neil, "Maintaining Nuclear Stability in South Asia," *Adelphi Paper 312*, 1997.

Krepon, Michael, "The Stability-Instability Paradox," *Prospects for Peace in South Asia* (Palo Alto, CA: Stanford University Press, 2004).

Levi, Michael, *On Nuclear Terrorism* (Cambridge, MA: Harvard University Press, 2007).

Lowther, Adam, *Challenging Nuclear Abolition* (Maxwell AFB, AL: Air Force Research Institute Papers, 2009).

Nuclear Threat Initiative, "Nuclear Weapons Budget Overview," www.nti.org/analysis/articles/us-nuclear-weapons-budget-overview/.

Obama, Barack H., address, Prague, Czechoslovakia, April 5, 2009.

Perkovich, George, *Do Unto Others* (Washington, DC: Carnegie Endowment for International Peace, 2013).

3 Does one size fit all?

Paul D. Schumacher

Introduction

The Roman motto, of *si vis pacem, para bellum* (if you wish for peace, prepare for war) highlights one early state's use of military threats to coerce another nation's behavior.[1] Not only states but individuals as well use the concept of deterrence on a daily basis. Almost every parent engages in the full spectrum of deterrence at some point when raising their children via rewards and threats to coerce desired behavior or prohibit undesired behavior.

With the detonation of two atomic bombs over Japan in August 1945, and the Soviet Union testing an atomic bomb in 1949, the academic field of nuclear deterrence began. Within the new field two theorists gained early prominence; Nobel laureate Thomas Schelling and Herman Kahn developed different analytical frameworks regarding rational adversary behavior during the Cold War.[2] Schelling proposed a stable balance of terror focused on mutually assured destruction emphasizing the element of chance.[3] Kahn emphasized an asymmetric balance of terror in favor of the United States and accompanied by defensive efforts.[4] Both theories focused on the ability of rational actors to deduce the inner workings of deterrence. While the overall concept of deterrence appears straightforward, according to the Department of Defense definition of manipulating behavior via threats, the actual process of deterrence involves highly complex, mental mechanisms and a variety of situational factors. The salient point for both theories concerns their development during a time when the dominant feature of the international strategic environment was a bipolar adversarial relationship between the nuclear-armed countries of the United States and the Soviet Union. The current strategic environment is no longer congruent with bipolarity.

With the end of the Cold War, the Soviet–American, adversarial relationship dissolved. The strategic environment that emerged looks markedly different and more complex with nine nuclear weapon states in addition to rogue states and violent non-state actors exploiting the proliferation of weapon technology.[5] To properly elucidate the growth in complexity, Figure 3.1 and Figure 3.2 are approximate depictions of the nuclear environment during the Cold War and today.

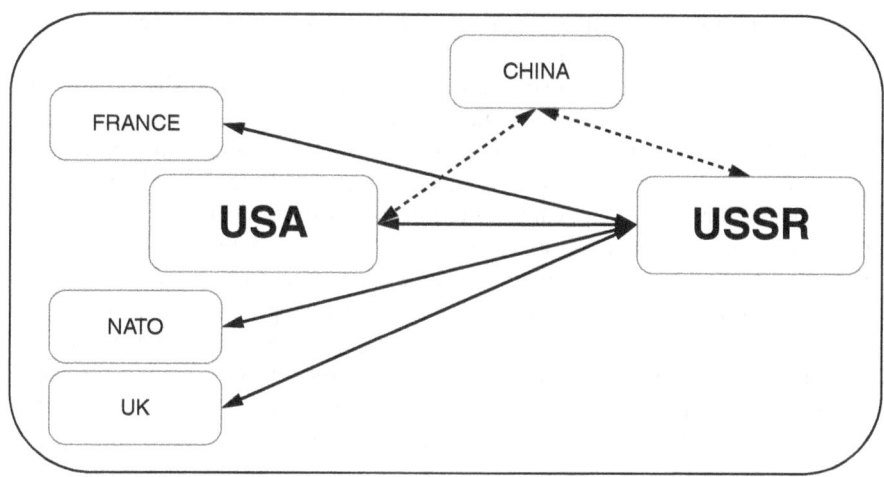

Figure 3.1 Graphic representation of the Cold War nuclear security environment through the early 1970s

Note
Solid line represents primary deterrence effort. Positioning represents alliance.

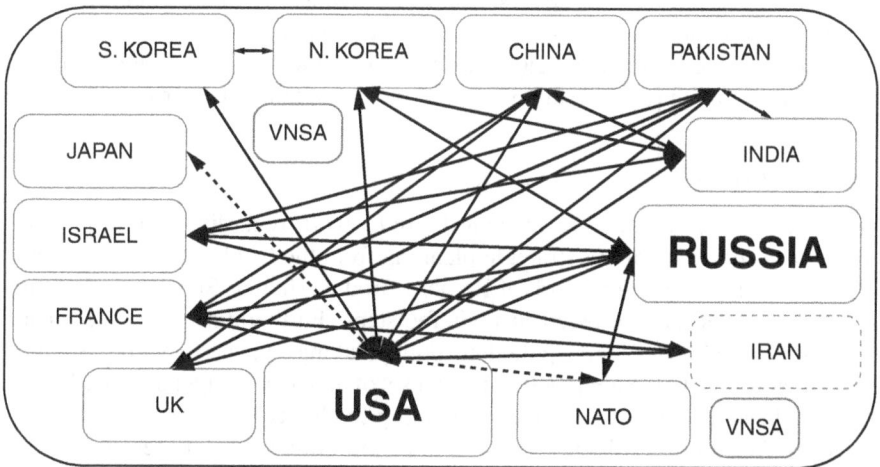

Figure 3.2 Graphic representation of the international nuclear security environment through 2010

Note
Dotted line represents extended deterrence agreements. Dashed line around Iran indicates suspected proliferation efforts. VNSA represents violent non-state actors.

Clearly, the increase in the number of actors, all with different cultural backgrounds and regional influences, create a more volatile, complex, uncertain, and ambiguous security environment.

The failure of deterrence to stem the increase in actors since the end of the Cold War leads to a debate regarding the utility of deterrence.[6] There exists some validity to this argument. The US, as the single greatest power, is in possession of a clearly superior conventional force and the largest strategic nuclear arsenal on the planet. But the US appears limited in its ability to coerce or deter other nation-states and non-state actors.[7] Instead of questioning the utility of deterrence, perhaps new strategic thinking on the utilization or tailoring of deterrence for the new strategic environment is required. This chapter examines which is more appropriate for today's environment: a general nuclear deterrence policy or a tailored deterrence policy. Given the inherent complexities of today's complex and multi-polar world, a tailored deterrence policy provides the best path toward credible deterrence success and promoting global stability and security.

What is wrong with tailored deterrence?

If tailored deterrence provides the best path forward to ensure global security and stability, one may rightly question why the US has not embraced this shift in deterrence thinking. There are three main intellectual and theoretical challenges to tailored deterrence. Classic deterrence, structural realism, and pragmatism each present potent arguments against embracing the concept of tailoring deterrence. A presentation of each argument includes an analysis of the current strategic environment depicted in Figure 3.2—supporting the validity of tailored deterrence.

Classic deterrence

While there currently exists a growing body of work regarding tailored deterrence, the cornerstone of US strategic planning from the Cold War to the present is focused on the theoretical work of Bernard Brodie, Thomas Schelling, Herman Kahn, and others that followed in their footsteps. Furthermore, the theoretical frameworks established and applied to policymaking appear to work. Simply stated, the Cold War ended without the use of nuclear weapons. Schelling's "stable balance of terror that leaves something to chance" for the rational actor served as the guiding principle for US strategic policy.[8] Kahn's emphasis on defensive capabilities with an asymmetric advantage in favor of the US arguably served as the basis for the Strategic Defense Initiative and current missile defense initiatives.[9]

The language and methodology used in nuclear policy documents remains relatively unchanged from the Cold War to the post-Cold War.[10] The pertinent question asks whether the changing security environment, from bipolar to multi-polar, diminishes the utility of classical deterrence tenets. And does the current language of deterrence and calculations of weapon requirements reflect the new environment or institutional inertia? Arguably the assumptions of Cold War deterrence no longer apply due to changes in the strategic environment and technological developments, however, institutional inertia may serve as the driving factor for the current approach to deterrence.[11]

The rationale proved largely valid, although evidence exists that the Soviet Union did not hold the same convictions.[12] Looking at Figure 3.2 and taking into account the varying arsenal size of the nuclear powers, it becomes harder to argue that all nations possess a secure second-strike capability. Thus, a major tenet of classical deterrence comes into question. Further, as the numbers of leaders in possession of nuclear weapons increase, the odds of all leaders behaving rationally decrease. As evidence suggests, in times of crisis, individuals begin to operate less rationally and rely more on heuristics to make decisions.[13]

The other complicating factor involves unintentional signaling. As seen in the case of India and Pakistan, recent dialogue with India over its nuclear weapons enterprise created negative signals in Pakistan potentially inducing new momentum for a buildup in Pakistani weapons.[14] Operating from a one-size-fits-all approach in today's multipolar world introduces greater potential for miscalculation and unintentional signaling due to greater complexity in the international security environment. Tailoring messages and capabilities to a specific leader in a specific context provides a higher degree of probability for deterrent success and lowers the probability of miscalculation and inadvertent signaling.

Structural realism

Structural realism, a variant of classical realism, argues that the tripartite concept of functional differentiation, ordering principles, and power distribution forces leaders to act in a specific manner regardless of their personal nature.[15] Kenneth Waltz, considered by many as the father of structural realism, presents a powerful argument that nuclear weapons increase stability in the international arena and that the weapon's magnitude of destruction affects all leaders in the same manner.[16] Regardless of the adversary, nuclear weapons serve a primary deterrence role.[17] This presents a powerful argument and the historical record provides some validation of Waltz's premises. The foundation of the structural realist argument involves a focus on great powers, since these states dominate the international system. According to Waltz, "A state becomes a great power not by military or economic capability alone but by combining political, social, economic, military, and geographic assets in more effective ways than other states can."[18]

Using this definition, structural realists would likely view the Cold War world the same as Figure 3.1 depicts, a bipolar environment. However, disagreement may occur with Figure 3.2 depicting a multipolar world. Arguably the situation, from a structural realist viewpoint, more closely resembles a unipolar world with the US as the sole superpower. For structural realists, a bipolar world—the Cold War—provides the best stability in the international system, followed by a unipolar world, and lastly a multipolar world creating the most instability and greatest probability for miscalculation.[19]

The fundamental argument then becomes whether the description of the international system today is one of unipolarity or multipolarity. Clearly, American capabilities create a scenario where there exists no peer competitor; however, US interactions with nuclear states create a perception of multipolarity. The most obvious case in favor of this perception involves the Bush administration's diplomatic actions against North Korea and conventional actions against Iraq.[20] Both countries are viewed as rogue states, with North Korea presenting a more imminent proliferation concern, and should dictate actions other than diplomacy from a realist perspective.[21] However, with China, a sponsor of North Korea, possessing a capable nuclear arsenal, a softer approach was necessary than with Iraq. Therefore, while relative power alone might indicate a unipolar international system, eight other nuclear weapon states arguably create a multipolar environment. Thus, tailoring American deterrence to reduce miscalculation and miscommunication could provide a better path forward.

Pragmatists

The argument against tailored deterrence from a pragmatic point of view does not necessarily encompass a single, coherent worldview regarding international relations. Realists, liberals, and constructivists may align with this simple and compelling argument: there exist nearly insurmountable obstacles to executing tailored deterrence.[22] The obstacles run the gamut from institutional bureaucracy to cognitive biases. Sean Larkin and David Yost offer poignant and detailed challenges to tailored deterrence from American and NATO perspectives, respectively.[23] While these authors posit a variety of challenges, their arguments overlap on one critical element of tailored deterrence—understanding the adversary.[24] The need to understand the adversary originates from ancient times, with history illustrating this need with both successful outcomes based on understanding the adversary and unsuccessful outcomes due to ignorance.[25] Larkin emphasizes the insurmountable aspect of this challenge pointing out the various heuristic and cognitive biases that *all* people use when making decisions.[26] Clearly, when applying this argument against the structure described in Figure 3.2, an already complex picture increases in complexity.

Should the US then reject the concept of tailored deterrence for a general deterrence approach since following the tailored deterrence path promises ever higher degrees of complexity and is based upon the foundation of understanding specific adversaries? The short answer is no. Sir Francis Bacon wrote, "*ipsa scientia potestas est*" (knowledge is power) in 1597 and much of deterrence theory builds on this foundation.[27] Regardless of which wave of deterrence theory one supports, a requirement exists to have some level of knowledge of the adversary to generate a cost-benefit calculation. The premise of tailored deterrence attempts to increase knowledge of a specific adversary, enabling a country to choose the best option to influence an adversary's cost-benefit calculation.

The complexity of the current international system reinforces the need to focus greater energy on different adversaries and determine how to best communicate clear deterrence messages. Will this approach guarantee desired results in all scenarios? The answer to this question remains the same for classical deterrence. How does one prove the outcome of a mental cost-benefit calculation? A commitment to tailored deterrence ensures the US expends effort attempting to better understand adversaries in today's environment versus relying on Cold War theories created in a vastly different security environment toward today's problems.

What is tailored deterrence?

Chronological development

With the end of the Cold War, the role of deterrence faded from its central role in guiding US national security. Multiple explanations describe this decline. Therese Delpech argues that the success of deterrence in the Cold War led to a loss of prominence.[28] Keith Payne suggests intellectual hubris: we think we know all that we need to know regarding deterrence.[29] Lawrence Freedman argues that changing circumstances reduced the need to rely on deterrence.[30] Regardless of which view accurately describes the reason, there exists wide agreement that intellectual thinking on deterrence theory declined precipitously after the Cold War.

The degree of detriment to national security resulting from the decline in intellectual thinking on deterrence remains a topic of debate. To address this decline in thinking, Payne (2001) introduced a detailed framework for the concept of tailoring deterrence to specific actors in specific situations to address a radically different international security environment.[31] He listed many factors for shifting from a Cold War deterrence framework to a tailored approach. One of the cornerstone issues involves intertwining the separate ideas of rational and reasonable behavior.[32] As stated earlier, the foundation of deterrence theory assumes a rational actor. Anticipation of how a potential adversary might respond to deterrent signals largely involves the process of mirror imaging and deductive logic, although the US expended great effort and cost to understand its Soviet adversary.[33] The fundamental problem of mirror imaging and deductive logic involves the introduction of inherent miscalculation and misperception errors due to personal cultural values or filters becoming imposed on adversaries.

To avoid introducing these systematic errors and to reduce the probability of miscalculation and misperception, Payne advocates a six-step information-gathering and analysis framework centered on the key decision makers in specific situations to best guide US efforts in tailoring deterrence capabilities.[34] For deterrence to work in a complex environment, policymakers need to cease expecting a rational and predictable adversary and begin gathering as much information about specific adversary characteristics, beliefs, and values as possible.[35] In Payne's words,

> The primary areas of interest in this framework are characteristics of: the pertinent leaderships/countries, their motivations, goals, and determination,

the nature of decision-making, the object of the friction (the "stakes" involved), the regional political/security context, and the sources of power available to the participants.[36]

The design of the framework gets inside the decision-making process, ascertains values, and identifies critical decision factors to better ensure the success of deterrence actions.[37] The framework serves as an empirical guide not as a magic bullet since no parsimonious solution exists in ensuring deterrence works.[38] More important, embracing the concept of tailoring deterrence breaks with the intellectual hubris of assuming that the Cold War deterrence framework knows how deterrence works in all situations.[39]

The first official government usage of tailored deterrence occurred in the 2006 *Quadrennial Defense Report* (QDR) followed by the 2006 *National Security Strategy* (NSS).[40] The 2006 QDR and NSS advocated tailoring deterrence and capabilities, yet failed to adequately describe the actual mechanics.[41] The 2010 NSS continues advocating tailoring deterrence and capabilities in the same manner as the 2006 edition.[42] To address the actual mechanics of tailored deterrence, the Department of Defense (DoD) published the Deterrence Operations Joint Operating Concept (DO-JOC) in 2006.[43] The DO-JOC serves as the single source document outlining the whole-of-government approach toward tailored deterrence. Larkin, in his critique of tailored deterrence, gives a thorough and penetrating review regarding the theoretical basis and mechanics of tailored deterrence contained within the DO-JOC.[44] His analysis that the DO-JOC "is an amalgam of second and third-wave deterrence theory, heavily influenced by effects-based operations concepts" correctly summarizes the theoretical underpinnings of the document.[45]

M. Elaine Bunn (2007) further developed the tailored deterrence concept by emphasizing the three key facets of tailoring involving actors, capabilities, and communication.[46] In tailoring to specific actors, she identifies the types of information the US needs to gather with the understanding that some information may be difficult to discern.[47] Regarding capabilities, Bunn points out the confusion in determining how to program and project capabilities needed for scenarios both in peacetime and crisis situations.[48] The final point highlights the need to tailor messages both verbal and non-verbal and avoid the sending of conflicting messages.[49] Bunn effectively highlights that everything the US says, does, and possesses sends verbal and non-verbal messages to potential adversaries.[50] The deeper meaning implied is that peacetime actions may have more impact on an adversary's perceptions than messages sent during a crisis.[51] A thorough understanding of this point would improve intentional messaging of American policymakers.

Jasen Castillo (2007) outlined the categories of adversary characteristics that influence tailored deterrence capability, credibility, and communication.[52] He highlighted potential differences, motivations, and dangers for near-peer competitors, rogue states, and non-state actors that motivate their use of unrestricted warfare and how these differences affect the implementation of tailored deterrence options.[53]

In 2009, Jeffrey Lantis examined the relationship between strategic culture and tailored deterrence. He explored links between culture and deterrence, identifying scope conditions that increase the utility of models for military-security policy.[54] These scope conditions included states with dominant cultural narratives, determined leadership, and prominent military organizations and identified potential areas where cultural knowledge helps explain patterns of non-state actors.[55]

Also in 2009, Kevin Murphy undertook the task of defining and developing an analytical framework that "helps identify, from the perspectives of the social and behavioral sciences, which questions analysts should consider asking of this research community in developing and evaluating tailored deterrence strategies."[56] The report also considers the challenges in synthesizing the information for decision makers when considering the relevance, comprehensiveness, and reliability of sources.[57]

As this time line shows, tailored deterrence represents a relatively recent approach towards the larger concept and more well developed concept of deterrence. While this time line does not represent an exhaustive list of all writings, it highlights the evolution of thinking regarding tailored deterrence and demonstrates the systematic development of the concept.

What tailored deterrence gets right

The factors and framework advocated by Payne and Bunn build from general deterrence theory by narrowing the scope for specific actors and situations and gaining as much insight as possible regarding adversary thinking, behavior, and values. The concept abandons the one-size-fits-all approach, based primarily on the nuclear deterrent. Instead, it opts for a whole-of-government approach utilizing all the instruments of power as deemed applicable based on the information and analysis available. While detractors argue against the merits of tailored deterrence and proponents acknowledge the difficulties required in following this approach, tailored deterrence is correct on many things. Three factors that tend to get lost in the debate over tailored deterrence deserve highlighting: (1) the focus on a specific adversary; (2) rational does not equal reasonable; and (3) the complex strategic environment as a game changer.

Focus on the specific adversary

While general deterrence theory acknowledges that in order to deter an adversary the adversary's perception of the deterrence threat must be credible and force a cost-benefit calculation, the theory then created the rational and predictable actor to develop a workable model. Adversary actions now become a mirror-imaging exercise, with possible courses of action tainted by Western cultural values. In contrast, tailored deterrence focuses on the specific adversary in its environment and analyzes the adversary's statements and actions to gain insight into the thought processes and decision-making calculus. Instead of mirror imaging behavior, tailored deterrence seeks to truly understand an adversary's behavior and perceptions. Tailoring seeks to get into the mind of the adversary to the greatest extent possible.[58] As the old refrain says, "the enemy gets a vote." Tailored deterrence seeks to predict the enemy vote in a realistic manner.[59]

Returning to Figure 3.2 as a depiction of today's complex strategic environment, the task tailored deterrence sets for itself is daunting. Multiple and varied actors exist in different regional contexts with varying interests and associations. With certainty, a temptation exists to simplify the complexity by assuming rational behavior and developing general policy options to deter or dissuade adversaries. However, the wide spectrum of cultural values, relationships, and American commitments, compounded by non-state actors and globalized communications, practically ensures that actions in one region affects other regions, and multiple interpretations and perceptions result. Without devoting the effort to understanding each adversary, a blanket approach logically leads to counterproductive and potentially detrimental follow-on effects. Attempting to understand a specific enemy's regional context, values, and beliefs logically leads to a better probability of successful dissuasion or deterrence with less probability of detrimental follow-on effects. Again, without complete knowledge no theory of deterrence can guarantee success in every situation; however, departing from an assumption of rationality, as defined in Western thought, provides a better foundation from which to navigate today's complex strategic environment.

Rational versus reasonable

The second area that tailored deterrence gets right concerns the separation between the meanings of rational and reasonable. While rational and reasonable are nearly synonymous in definition their usage possesses differences that can lead to erroneous results. Rational implies cold logic in decision making while reasonable incorporates norms and behavior, as understood by the recipient. A rational individual can make demands that appear unreasonable. Tailored deterrence acknowledges that all individuals possess filters which modify rational, and more importantly, reasonable thinking. According to Payne, Cold War deterrence thinking ignores the opponent's filter and predicts opponent behavior using our own filter.[60] The tailored deterrence framework seeks to understand the adversary's filter by examining the psychological and cultural anthropological factors that create the filter. Understanding the filter of an adversary better equips policymakers and analysts in predicting the interpretations and perceptions of American actions. As Bunn writes, "The message intended by our actions and statements is less important than the message received."[61] A poignant example of this statement occurred in the Truman administration when the Secretary of State's comments were interpreted as excluding South Korea from a statement of American interests, thus potentially opening the way for a North Korea invasion.[62] With certainty, the US did not intend to send such a signal but failure to understand how North Korea and the Soviet Union would perceive the statement arguably opened the way.

Again, viewing today's complex environment, the same temptation to simplify applies with the same logical conclusions. Reasonable from a US point of view does not equal reasonable in every other region. The different psychological and cultural anthropological factors clearly lead to different interpretations of rational and reasonable thinking and actions. A blanket approach establishes the conditions for miscalculation and misperception. Making the attempt to understand the filters allows the US to tailor capabilities and messages that produce a better probability of success in assurance of allies and deterrence of adversaries. The increase in complexity and risk of the current security environment with multiple nuclear-armed actors practically demands that the US seek to understand these adversary filters to avoid misperception and miscalculation.

The game has changed

The risks and potential catastrophic consequences of failed nuclear deterrence require serious thought and deliberation. The utilization of a game analogy in the following argument is meant only as an analogy for illustrative purposes. Arguably, the fundamental question in applying nuclear deterrence concerns the determination of qualitative and quantitative requirements for providing assurance and deterrence. In modern lexicon, how much is enough? Payne makes the compelling argument that Schelling's stable balance of terror took prominence since it allowed for an empirical calculation of warheads to targets.[63] Post-Cold War thinking on the arsenal reflects a continuation of the calculation process.[64] The problem with this approach in today's complex environment is that a stable balance of terror paradigm ceases to exist. No longer does the US play a chess match against a singular, evenly matched opponent. Nine asymmetrical players now sit at the proverbial poker table with others attempting to join. Furthermore, the stakes and rules in the game are different for each player. The shift towards tailored deterrence acknowledges the changing game and attempts to understand the internal dynamics of the game. General deterrence inherently assumes the other players abide by a set of understood rules with no intent of breaking them. The inherent risks, given the consequences of failure, are too high to utilize a one-size-fits-all approach. It is better to seek to understand the personalities, idiosyncrasies, and risk tolerance of each player. This allows for improved prediction of which "cards" to play and in what order to best influence the desired outcome.

Recommendations

As the 2006 and 2010 NSS demonstrate, the US intends to embrace a tailored, whole-of-government, approach towards assurance and deterrence. The framework and guidelines established to date serve as a good foundation; however, the requirement exists for more intellectual thinking and honest debate on the topic. In reviewing an abundance of literature regarding deterrence in the twenty-first century, two areas stand out as potential areas needing further examination:

modernization of nuclear weapons and the exploration of behavioral economics as an explanatory tool.

Revitalizing nuclear weapons

According to Stephen Younger, nuclear weapons differ in design based upon the intended mission and technological capabilities present at the time of construction.[65] The current US arsenal designed and built during the Cold War for the Soviet Union held specific target sets at risk. In the post-Cold War strategic environment American thinking on deterrence shows change and adaptation to this new environment; however, the weapons and infrastructure, while smaller, remain relatively unchanged. With the monumental advances in technology, our current arsenal truly remains a technological relic of the Cold War. Tailoring capabilities to specific adversaries requires adaptable capabilities to the spectrum of targets. While US conventional capabilities display exceptional adaptability and periodic modernization, the nuclear arsenal needs modernization to include new weapons that will hold the entire spectrum of targets at risk.

Another argument in favor of revitalization of nuclear deterrence involves US credibility. Actions speak louder than words. While American rhetoric appears credible, the decaying infrastructure sends a signal to adversaries that the US considers nuclear weapons relics of the Cold War and arguably has little will to actually use a nuclear weapon again. Revitalization sends a clear and credible signal that the United States remains committed. As credibility provides the foundation for all deterrence efforts, modernization needs to occur.

Behavioral economics

In 1979, Amos Tversky and Daniel Kahneman developed a behavioral economics model called prospect theory that resulted in the 2002 Nobel Prize in Economics.[66] Prospect theory and the modified version of cumulative prospect theory look at decision making involving risk and uncertainty, incorporating cognitive biases and heuristics.[67] While the third wave of deterrence theory involves cognitive biases, no clear connection of prospect theory to deterrence currently exists.[68] Currently, prospect theory suggests more decision phenomena than established decision-making theories can explain. As a simplified example, Adam Lowther makes the argument that possessing nuclear weapons serves as a stabilizing influence.[69] This argument fits with the prediction of prospect theory that as individuals gain value (weapons) their reference point shifts and their view of gains and losses can lead to risk-adverse behavior.[70] The preceding example is meant only to show the potential correlation between observed state behavior and prospect theory's prediction and does not imply a de facto correlation. Much more rigorous analysis would be needed to determine if a true correlation exists. However, given the seminal work between game theory and deterrence, and the fact that prospect theory attempts to explain an actor's behavior which tailored deterrence also seeks to understand, a clear opportunity exists to explore the potential linkages and potential correlations.

Conclusion

With the end of the Cold War, innovative thinking on deterrence diminished and the resulting international security environment grew in complexity as additional nuclear-armed adversaries and violent non-state actors came to threaten American interests. In response to changes in the environment, the US changed directions towards its approach regarding deterrence, shifting from a one-size-fits-all mentality to tailoring deterrence messages and capabilities for specific actors in specific situations. The shift resulted in varied degrees of acceptance and criticism. The new framework seeks to understand the norms, values, culture, and decision-making processes of adversaries within their regional and individual contexts to better predict adversary behavior and avoid misperception and miscalculation that undermine deterrence efforts. Tailored deterrence shifts from the Cold War deterrence framework that utilized mirror imaging and rational deductive logic and seeks to clearly understand how adversaries filter and process incoming information. Additionally, tailored deterrence views the world as it exists today with multiple nuclear-armed states and stakes in the game that clearly changes the playing field.

Notes

1 Lawrence Freedman, *Deterrence* (Cambridge, UK: Polity Press, 2004), 7.
2 Keith B. Payne, *The Great American Gamble: Deterrence Theory and Practice from the Cold War to the Twenty-First Century* (Fairfax, VA: National Institute of Public Press, 2008), 29–32.
3 Thomas Schelling, *The Strategy of Conflict* (Cambridge, MA: Harvard University Press, 1960), 207–229.
4 Herman Kahn, *On Thermonuclear War* (Princeton, NJ: Princeton University Press, 1960), 213.
5 Barry Schneider and Patrick Ellis, *Tailored Deterrence: Influencing States and Groups of Concern*, 2nd ed. (Maxwell AFB, AL: USAF Counterproliferation Center, 2012), 3.
6 Naval Studies Board, *Post-Cold War Conflict Deterrence* (Washington, DC: National Academy Press, 1997), 113, www.nap.edu/openbook.php?record_id=5464.
7 White House, *National Security Strategy* (Washington, DC: White House, May 2010), 15.
8 Payne, *The Great American Gamble*, 31.
9 Kahn, *On Thermonuclear War*, 213 and 557.
10 George W. Bush, *National Security Strategy* (Washington, DC: White House, 2006); and Barack Obama, *National Security Strategy* (Washington, DC: White House, 2010).
11 Keir A. Lieber and Daryl G. Press, "The New Era of Nuclear Weapons, Deterrence, and Conflict," *Strategic Studies Quarterly* 7, 1 (Spring 2013): 4.
12 Keith B. Payne, *The Fallacies of Cold War Deterrence and a New Direction* (Lexington, KY: University Press of Kentucky, 2001), 19–27.
13 Daniel Kahneman, *Thinking Fast and Slow* (New York: Farrar, Straus and Giroux, 2011), 7.
14 Paul K. Kerr and Mary B. Nitikin, "Pakistan's Nuclear Weapons: Proliferation and Security Issues," Congressional Research Service, *Report for Congress*, RL34248, May 2009, 4, www.crs.gov.

15 James W. Forsyth Jr., "The Common Sense of Small Nuclear Arsenals," *Strategic Studies Quarterly* (Summer 2012): 93–108.
16 Kenneth Waltz, "The Spread of Nuclear Weapons: More May Be Better," *Adelphi Papers*, 171 (1981): 6, https://www.mtholyoke.edu/acad/intrel/waltz1.htm.
17 Keir A. Lieber and Daryl G. Press, "The Nukes We Need: Preserving the American Deterrent," *Foreign Affairs* 88, 6 (November/December 2009): 41, www.metu.edu. tr/~utuba/Lieber-Press.pdf.
18 Waltz, "The Spread of Nuclear Weapons," 4.
19 Waltz, "The Spread of Nuclear Weapons," 2.
20 Jaechun Kim and David Hundt, "US Policy toward Rogue States: Comparing the Bush Administration's Policy toward Iraq and North Korea," *Asian Perspective*, 35 (2011): 241, http://search.proquest.com.aufric.idm.oclc.org/docview/928083865/fullte xtPDF/13F53370DD55AF8E4C3/4?accountid=4332.
21 Kim and Hundt, "US Policy toward Rogue States," 245.
22 Sean Larkin, *Cracks in the New Jar: The Limits of Tailored Deterrence* (Carlisle, PA: Army War College, 2011), 1; and David Yost, "NATO and Tailored Deterrence: Surveying the Challenges," World Security Network, May 2009, www.worldsecuri- tynetwork.com/Other/Yost-David/NATO-and-Tailored-Deterrence-Surveying-the- Challenges.
23 Larkin, *Cracks in the New Jar*, 6.
24 Larkin, *Cracks in the New Jar*, 14.
25 Yost, "NATO and Tailored Deterrence."
26 Larkin, *Cracks in the New Jar*, 14.
27 Sir Francis Bacon, *Novum Organum, Part I, Aphorism III* (Boston, MA: Taggard & Thompson, 1863, volume VIII), 67–68.
28 Therese Delpech, *Nuclear Deterrence in the 21st Century: Lessons from the Cold War for a New Era of Strategic Piracy* (Santa Monica, CA: RAND Corporation, 2012), 10.
29 Payne, *Fallacies of Cold War Deterrence*, 99.
30 Freedman, *Deterrence*, 21.
31 Payne, *Fallacies of Cold War Deterrence*, 104–114.
32 Payne, *Fallacies of Cold War Deterrence*, 87.
33 Payne, *Great American Gamble*, 304; and M. Elaine Bunn, "Can Deterrence Be Tai- lored?" *Strategic Forum* 225 (January 2007): 3.
34 Payne, *Fallacies of Cold War Deterrence*, 104–114.
35 Payne, *Fallacies of Cold War Deterrence*, 102.
36 Payne, *Fallacies of Cold War Deterrence*, 103.
37 Bunn, "Can Deterrence Be Tailored?" 2.
38 Payne, *Fallacies of Cold War Deterrence*, 110.
39 Payne, *Great American Gamble*, 249.
40 US Department of Defense, *Quadrennial Defense Review Report* (Washington, DC: US Department of Defense, February 6, 2006), 49, www.defense.gov/qdr/report/ report20060203.pdf; and White House, *National Security Strategy*, 43.
41 Larkin, *Cracks in the New Jar*, 8.
42 Larkin, *Cracks in the New Jar*, 9.
43 US Department of Defense, *Deterrence Operations Joint Operating Concept Version 2.0*, (Washington, DC: US Department of Defense, December 2006), 5, www.dtic. mil/futurejointwarfare/concepts/do_joc_v20.doc.
44 Larkin, *Cracks in the New Jar*, 8–11.
45 Larkin, *Cracks in the New Jar*, 11.
46 Bunn, "Can Deterrence Be Tailored?" 1.
47 Bunn, "Can Deterrence Be Tailored?" 3.
48 Bunn, "Can Deterrence Be Tailored?" 5.
49 Bunn, "Can Deterrence Be Tailored?" 7.

50 Bunn, "Can Deterrence Be Tailored?" 7.
51 Bunn, "Can Deterrence Be Tailored?" 7.
52 Jasen Castillo, "Tailored Dissuasion and Deterrence?" Unrestricted Warfare Sympo-
    sium Proceedings, Washington, DC, 2007, 177, www.jhuapl.edu/urw_symposium/
    proceedings/2007/papers/Castillo.pdf.
53 Castillo, "Tailored Dissuasion and Deterrence?" 179.
54 Jeffrey S. Lantis, "Strategic Culture and Tailored Deterrence: Bridging the Gap
    between Theory and Practice," *Contemporary Security Policy* 30, 3 (December 2009):
    467, www.contemporarysecuritypolicy.org/assets/CSP-30-3-Lantis.pdf.
55 Lantis, "Strategic Culture and Tailored Deterrence," 468.
56 Kevin R. Murphy, *Defining an Analytic Framework for Tailored Deterrence: Con-
    tributions of Social Science Research: Initial Steps in Defining an Analytical
    Framework* (Ft. Belvoir, VA: Defense Threat Reduction Agency, Advanced Systems
    and Concepts Office, Report Number ASCO2009–008,2009), 4, www.hsdl.org/
    ?view&did=716139.
57 Murphy, *Defining an Analytic Framework for Tailored Deterrence*, 5.
58 Bunn, "Can Deterrence Be Tailored?" 3.
59 Jonah Goldberg, "Our Enemies Get a Vote," *National Review Online*, May 2013,
    www.nationalreview.com/article/349499/our-enemies-get-vote#!.
60 Payne, *Great American Gamble*, 239.
61 Bunn, "Can Deterrence Be Tailored?" 7.
62 Dean Acheson, Secretary of State, remarks made to the National Press Club Speech,
    January 12, 1950, Truman Library Archives, www.trumanlibrary.org/whistlestop/
    study_collections/korea/large/documents/pdfs/kr-3-13.pdf.
63 Payne, *Great American Gamble*, 100.
64 Payne, *Great American Gamble*, 238.
65 Stephen M. Younger, *The Bomb: A New History* (New York: Harper Collins,
    2009), 96.
66 Kahneman, *Thinking Fast and Slow*, 14.
67 Kahneman, *Thinking Fast and Slow*, 286.
68 Larkin, *Cracks in the New Jar*, 6.
69 Adam Lowther, "Iran's Two-Edged Bomb," *New York Times*, February 9, 2010,
    www.nytimes.com/2010/02/09/opinion/09lowther.html?_r=0.
70 Kahneman, *Thinking Fast and Slow*, 334.

Bibliography

Acheson, Dean, Secretary of State, remarks made to the National Press Club Speech,
    January 12, 1950, Truman Library Archives, www.trumanlibrary.org/whistlestop/
    study_collections/korea/large/documents/pdfs/kr-3-13.pdf.
Bacon, Sir Francis, *Novum Organum, Part I, Aphorism III* (Boston, MA: Taggard &
    Thompson, 1863, volume VIII).
Bunn, M. Elaine, "Can Deterrence Be Tailored?" *Strategic Forum* 225 (January 2007).
Bush, George W., *National Security Strategy* (Washington, DC: White House, 2006).
Castillo, Jasen, "Tailored Dissuasion and Deterrence?" Unrestricted Warfare Symposium
    Proceedings, Washington, DC, 2007, www.jhuapl.edu/urw_symposium/proceed-
    ings/2007/papers/Castillo.pdf.
Delpech, Therese, *Nuclear Deterrence in the 21st Century: Lessons from the Cold War
    for a New Era of Strategic Piracy* (Santa Monica, CA: RAND Corporation, 2012).
Forsyth Jr., James W., "The Common Sense of Small Nuclear Arsenals," *Strategic
    Studies Quarterly* (Summer 2012).
Freedman, Lawrence, *Deterrence* (Cambridge, UK: Polity Press, 2004).

Goldberg, Jonah, "Our Enemies Get a Vote," *National Review Online*, May 2013, www. nationalreview.com/article/349499/our-enemies-get-vote#!.

Kahn, Herman, *On Thermonuclear War* (Princeton, NJ: Princeton University Press, 1960).

Kahneman, Daniel, *Thinking Fast and Slow* (New York: Farrar, Straus and Giroux, 2011).

Kerr, Paul K., and Nitikin, Mary B., "Pakistan's Nuclear Weapons: Proliferation and Security Issues," Congressional Research Service, *Report for Congress*, RL34248, May 2009, www.crs.gov.

Kim, Jaechun, and Hundt, David, "US Policy toward Rogue States: Comparing the Bush Administration's Policy toward Iraq and North Korea," *Asian Perspective*, 35 (2011), http://search.proquest.com.aufric.idm.oclc.org/docview/928083865/fulltextPDF/13F53 370DD55AF8E4C3/4?accountid=4332.

Lantis, Jeffrey S., "Strategic Culture and Tailored Deterrence: Bridging the Gap between Theory and Practice," *Contemporary Security Policy* 30, 3 (December 2009), www. contemporarysecuritypolicy.org/assets/CSP-30-3-Lantis.pdf.

Larkin, Sean, *Cracks in the New Jar: The Limits of Tailored Deterrence* (Carlisle, PA: Army War College, 2011).

Lieber, Keir A., and Press, Daryl G., "The New Era of Nuclear Weapons, Deterrence, and Conflict," *Strategic Studies Quarterly* 7, 1 (Spring 2013).

Lieber, Keir A., and Press, Daryl G., "The Nukes We Need: Preserving the American Deterrent," *Foreign Affairs* 88, 6 (November/December 2009), www.metu.edu.tr/ ~utuba/Lieber-Press.pdf.

Lowther, Adam, "Iran's Two-Edged Bomb," *New York Times*, February 9, 2010, www. nytimes.com/2010/02/09/opinion/09lowther.html?_r=0.

Murphy, Kevin R., *Defining an Analytic Framework for Tailored Deterrence: Contributions of Social Science Research: Initial Steps in Defining an Analytical Framework* (Ft. Belvoir, VA: Defense Threat Reduction Agency, Advanced Systems and Concepts Office, Report Number ASCO 2009–008, 2009), www.hsdl.org/?view&did=716139.

Naval Studies Board, *Post-Cold War Conflict Deterrence* (Washington, DC: National Academy Press, 1997), www.nap.edu/openbook.php?record_id=5464.

Obama, Barack, *National Security Strategy* (Washington, DC: White House, 2010).

Payne, Keith B., *The Fallacies of Cold War Deterrence and a New Direction* (Lexington, KY: University Press of Kentucky, 2001).

Payne, Keith B., *The Great American Gamble: Deterrence Theory and Practice from the Cold War to the Twenty-First Century* (Fairfax, VA: National Institute of Public Press, 2008).

Schelling, Thomas, *The Strategy of Conflict* (Cambridge, MA: Harvard University Press, 1960).

Schneider, Barry, and Ellis, Patrick, *Tailored Deterrence: Influencing States and Groups of Concern*, 2nd ed. (Maxwell AFB, AL: USAF Counterproliferation Center, 2012).

US Department of Defense, *Deterrence Operations Joint Operating Concept Version 2.0* (Washington, DC: US Department of Defense, December 2006), www.dtic.mil/future-jointwarfare/concepts/do_joc_v20.doc.

US Department of Defense, *Quadrennial Defense Review Report* (Washington, DC: US Department of Defense, February 6, 2006), www.defense.gov/qdr/report/report2006 0203.pdf.

Waltz, Kenneth, "The Spread of Nuclear Weapons: More May Be Better," *Adelphi Papers*, 171 (1981), https://www.mtholyoke.edu/acad/intrel/waltz1.htm.

White House, *National Security Strategy* (Washington, DC: White House, May 2010).

Yost, David, "NATO and Tailored Deterrence: Surveying the Challenges," World Security Network, May 2009, www.worldsecuritynetwork.com/Other/Yost-David/NATO-and-Tailored-Deterrence-Surveying-the-Challenges.

Younger, Stephen M., *The Bomb: A New History* (New York: Harper Collins, 2009).

4 Nuclear budgeting

The perilous impact of uncertainty

Michaela Dodge and Adam B. Lowther

Introduction

Since the end of the Cold War, the United States has struggled to clearly articulate the purpose of its nuclear forces. Unprecedented reductions in its nuclear arsenal, restraints on nuclear weapons testing and new designs, and a shift in focus to more likely, albeit less dangerous, national security problems led to a physical and human capital deterioration within the nuclear weapons enterprise. An intellectual confusion regarding the purpose of nuclear weapons is mirrored in budgeting for the enterprise and associated defense programs. Federal officials and nuclear enterprise analysts do not agree on how much the US is actually spending on nuclear forces, making a straightforward analysis of nuclear enterprise-related expenditures less accurate than desired. This creates an opportunity to inflate nuclear weapons costs to advance a political agenda—while confusing the public concerning the cost of the arsenal.

With the nuclear weapons budget jointly prepared by the National Nuclear Security Administration (NNSA) and the Department of Defense (DoD), it is understandable that there is some confusion as to the total cost of nuclear weapons-related activities. In its current form, the process is guided by the president's priorities (usually set forth in the *Nuclear Posture Review* [NPR]), which includes a role for the Department of Defense, Department of Energy, the White House, and Congress. The responsibility for determining expenditures is jointly shared by the House and Senate Appropriations Committees and the House and Senate Armed Services Committees. In short, the process for determining fiscal expenditures on the nuclear enterprise is complex and difficult to follow.

The following chapter is not meant to be an exhaustive description of all elements of the nuclear weapons budget or the budget process. Rather, it will discuss the main points of disagreement over fiscal issues between nuclear weapons critics and supporters; clarify why it is difficult to reach widespread agreement, inside and outside government, on how much the nation spends on its nuclear arsenal; and provide illustrations of the existing complications of nuclear weapons budgeting.

Two broad issues are the main reason for disagreement regarding the costs of nuclear weapons-related expenditures. One is the dual nature of many programs within the nuclear enterprise. For example, what portions of a nuclear-capable bomber crew's recruitment, training, and maintenance should be included in the nuclear weapons budget when almost all bombers fly mostly conventional missions? How should the Departments of Energy and Defense account for the costs of nuclear science projects that may not only benefit the nation's nuclear weapons program but many other unrelated scientific efforts, such as the National Ignition Facility? Should the cost of advancing general science be counted toward the nuclear weapons budget? How about the cost of building national security science facilities—that work on nuclear weapons-related research most of the time? The correct answer to these and similar questions is at the heart of some disagreements between critics of the nuclear arsenal and its advocates.

The second issue that complicates nuclear weapons budgeting is a disagreement over the definition of nuclear weapons modernization. The US is not developing new nuclear warheads but is extending the service-lives of currently deployed warheads: W78 and W87 for the Minuteman intercontinental ballistic missile, W-76-0/1 and W88 for the D5 submarine-launched ballistic missile, the B61 family of nuclear warheads for the dual-capable aircraft and B-2/B-52 bombers, and the W-80-1 air-launched cruise missile for the B-52. Over the next few decades, the administration's stockpile plan is to maintain three interoperable ballistic missile warheads and two interoperable air carrier warheads—all while the US needs to recapitalize its nuclear weapons delivery platforms (bombers, intercontinental ballistic missiles, air-launched cruise missiles, long-range standoff weapons, and both the ballistic missile submarines (SSBN) and submarine-launched ballistic missiles they carry) over the next two decades. This requires planning and investing resources now—while making the nuclear posture decisions that will impact national security. A second point of contention between critics and advocates of the arsenal is whether life extension programs (LEPs) count toward nuclear weapons modernization even though no new nuclear weapons are produced. The answer to this question largely depends on whether one is a critic or an advocate of modernization programs.

Given the two distinctly different views held by critics and advocates of the nuclear arsenal, the remainder of this chapter attempts to illuminate some of these important issues by first offering the arguments advanced by critics, followed by the arguments advanced by advocates. The chapter concludes with a set of recommendations we believe will ensure the nation's nuclear deterrent capability will remain credible for many years to come.

Critics: nuclear weapons are too expensive

President Obama's 2009 Prague speech called for the goal of pursuing a world without nuclear weapons and re-energized the public debate concerning the role of nuclear weapons in US national security—reinvigorating advocates of nuclear disarmament.[1] A renewed interest in nuclear weapons and a weak economy, which increased pressures on the federal budget, led to greater efforts to quantify how much the US spends on its nuclear weapons and supporting infrastructure.

Over the past five years a number of reports and studies have examined fiscal expenditures on the nuclear enterprise in an effort to determine the nuclear arsenal's total cost. Such studies are primarily conducted by organizations that seek to reduce and, ultimately, eliminate nuclear weapons and are designed to lead readers to believe the nuclear arsenal is too costly to maintain. Four studies are worth a brief examination because they are the most comprehensive efforts to understand fiscal expenditures on the nuclear enterprise from end to end.

The earliest of these studies was the Carnegie Endowment for International Peace's examination of nuclear enterprise costs published in 2009. The study divided nuclear enterprise expenditures into five categories: nuclear forces and operational support; deferred environmental and health costs; missile defense; nuclear threat reduction; and nuclear incident management. The study estimates nuclear weapons and weapons-related programs to be "at least" $52.4 billion in fiscal year (FY) 2008.[2]

The study adopted a very inclusive methodology and generally does not distinguish between dual-capable systems and dual-purpose budget lines, which may diminish the accuracy of the study's finding.[3] However, the Carnegie study is worth mentioning because it sets the high end of all cost estimates for the nuclear enterprise.

In 2012, the Henry L. Stimson Center produced a study that examined the cost of the nuclear enterprise, estimating an annual cost of $32.6 billion.[4] The Stimson report provides a good overview of where the discrepancies among different nongovernmental nuclear-force cost estimates are and uses an inductive method to come to its own conclusions about the costs of nuclear forces. It provides one of the most comprehensive nuclear mission budget estimates available. The report identifies operating costs; research, development, testing and evaluation (RDT&E); command and control (C2); and other operations and support.[5]

According to the Stimson study, operating costs include all expenditures associated with nuclear-capable bombers, nuclear strike-affiliated commands, and some costs of defensive measures for an estimated total of $12.6 billion.[6] The report estimates the RDT&E category at $1.58 billion. Within the C2 category, satellite communications and early warning systems are estimated at about $5.6 billion for nuclear-strike forces. The other operations and support category includes: training and recruiting, medical, family housing, centralized supply and maintenance, and centralized administration, which add up to $4.1 billion. The study also identified $8.7 billion in NNSA costs it attributes to supporting the nuclear-strike mission.

The Stimson analysis includes all expenditures associated with the bomber leg of the triad and some costs of tankers and command-and-control architecture. The report's ten-year strategic nuclear forces projection includes the costs associated with a new nuclear-capable bomber and some expenditures for the acquisition of aerial refueling tankers.[7] The report estimates the modernization costs of strategic forces between $370.3 billion and $412.4 billion for FY 2013–2022.[8]

The Congressional Budget Office (CBO), which strives for unbiased analysis, released its own estimate of nuclear enterprise costs in December 2013.[9]

According to the CBO, the cost of US nuclear forces is $373.4 billion for 2014–2023 (adjusted for the CBO's estimate of historical cost growth), roughly $37 billion annually.[10] Other nuclear-related activities will consume $218 billion over the same time period.[11] Indeed, expenditures will continue beyond the next decade as the US continues to modernize its nuclear triad.

According to the CBO, the costs of the nuclear triad include operations (operations and maintenance as well as military personnel); sustainment (acquisition and sustainment costs); and modernization (LEPs and delivery-system replacement).[12] The CBO attributes 25 percent of the costs associated with the B-52H bomber and the new long-range strike bomber and 100 percent of the costs associated with the B-2 bomber to the nuclear mission.[13] Clarity regarding the assumptions made within the CBO report is one of its main strengths. However, as with all such studies, the data is an imperfect estimate that depends on what is included within the cost estimate. For example, the majority of B-2 operations are unrelated to the nuclear mission, which means 100 percent of B-2 costs should not be attributed to the nuclear mission.

In 2014 The James Martin Center for Nonproliferation Studies at the Monterey Institute of International Studies (MIIS) estimated the cost of "maintaining the current arsenal, buying replacement systems, and upgrading existing nuclear bombs and warheads" at about $1 trillion over the next 30 years—$33 billion annually.[14] The report selected a 30-year estimate to capture the costs associated with the delivery platforms modernization, a bulk of which occurs after the usual ten-year time frame that most reports cover. The report correctly points out that US nuclear posture decisions must be driven by a sound strategic debate rather than a budgetary choice. The MIIS study includes the costs of a new bomber and all the costs of the NNSA in its estimate.

The single greatest shortcoming of the study is its time horizon. Predicting nuclear enterprise costs for the next three decades is highly susceptible to events overcoming predictions. While reaching $1 trillion of predicted outlays makes for sensational headlines, it is unlikely to be accurate and may be high or low.

With the exception of the CBO report, each of the studies argues, to a lesser or greater degree, that the cost of the nuclear arsenal is too high and/or unsustainable—necessitating a reduction in the size of the arsenal. How that reduction is best achieved may vary, but the main point of the critics is clear: spending between $32.6 billion (at the low end) and $57.2 billion (at the high end) is not the best use of federal dollars. The critics maintain that given the current challenges the United States faces with sequestration set to resume in FY 2016, and the growing threat of the Islamic State, spending 5 percent of the defense budget on the nuclear enterprise is difficult to justify.

The fiscal expense of the nuclear enterprise is not the only reason opponents advocate a reduction or elimination of the nuclear arsenal. Other arguments are addressed elsewhere, while this chapter focuses on fiscal issues. Thus, the following section turns to the arguments made by advocates of the nuclear arsenal.

Advocates: nuclear weapons are affordable and essential

The main danger associated with nuclear weapons budget estimates, like those highlighted in the previous section, is they often advance the notion that arbitrary budgeting decisions should play the most significant role in determining the nation's nuclear posture. In the current fiscal environment "it is impossible to make a strategic decision," according to General Charles Jacoby Jr., Commander of the North American Aerospace Defense Command and United States Northern Command.[15] Much of the discussion surrounding the fiscal costs of the nuclear enterprise are occurring within a strategic vacuum where federal expenditures on nuclear weapon and delivery systems are presented absent any context—strategic or fiscal.

While it is true fiscal resources are finite, sound strategic planning should start with an assessment of threats and capabilities. Based on these estimates, the United States should prepare a budget and, if needed, make trade-offs in a way that maximizes the chances of being successful in the most probable contingencies, while also hedging against worst case scenarios. This is an essential point because it is important for any nation to understand that the most likely and most deadly threats are not one and the same. So what budget assumptions must be made and what perspective must be taken regarding the overall federal budget?

First, the nuclear weapons budget must be considered within the context of the overall defense and federal budgets. In 2013, major entitlement programs consumed approximately 49 percent of federal spending.[16] Defense accounted for about 18 percent the same year. President Obama requested $495.6 billion for the DoD in his FY 2015 budget request.[17] Of the DoD's budget, the Air Force's $109.3 billion baseline budget for 2015 devotes approximately 5 percent ($5.5 billion) to the Intercontinental Ballistic Missile (ICBM) and bomber legs of the nuclear triad.[18] The SSBN leg of the triad accounts for approximately 6.7 percent ($10 billion) of the Navy's $148 billion baseline budget for 2015.[19]

In addition to DoD expenditures, the President's request for the NNSA, the agency in the Department of Energy responsible for nuclear warheads, was $11.658 billion.[20] The NNSA's budget request incorporates four different accounts: weapons activities ($8.315 billion); defense nuclear nonproliferation ($1.555 billion); naval reactors ($1.377 billion); and salaries and expenses ($411 million).[21] This brings the approximate, defense-specific total cost of the arsenal to approximately $27 billion for FY 2015. While such a sum is undeniably significant, when placed within the geostrategic context and the larger federal budget, arguments suggesting the nuclear arsenal is unaffordable are not valid. The simple fact is the United States must spend sufficiently to ensure its adversaries are deterred from attacking the United States, its allies, and its interests, all while assuring allies that the American nuclear umbrella remains credible. If providing such security were to cost a trillion dollars over the next 30 years ($33 billion annually), it would prove a worthwhile and minor investment when compared to many other expenditures of the federal government.

For example, the average American spends 10.3 percent of their annual income on insurance, while the federal government spends 0.0089 percent of the federal budget on nuclear weapons—the ultimate insurance policy.[22] With about 217,000,000 taxpayers in the United States, this means each taxpayer spends about $152 each year ($12 per month) on nuclear weapons and the security they provide.[23] For the value US nuclear weapons bring to US national security, the country cannot afford to allow the arsenal to decline in number, capability, or credibility.

What is often underappreciated is the fact that nuclear weapons are unique. The caution they generate in the mind of an adversary may be difficult to quantify and even harder to assign a monetary value, but individuals (average citizens, terrorists, and national leaders), regardless of nationality, instinctively know they are different. Not only do they have a sense that nuclear war is more destructive than conventional war, but they are more cautious in accepting risk when there is a potential for nuclear use.[24]

It is difficult for critics to argue that it is American spending on nuclear weapons that leads other states to pursue nuclear weapons. Other countries obtain them based on their own interests and perceived threats, not because the United States possesses them. It is worth noting that new nuclear weapon states emerged during a time when the United States and Russia were in the midst of large-scale nuclear arms reduction.[25] It is also important to understand that there is a fundamental asymmetry between the American nuclear arsenal and the arsenals of other states because while the United States seeks to target military targets, its leaders value life more than dictatorial and authoritarian regimes possessing nuclear weapons. Cities are inherently easier to destroy than military targets. One final point is instructive: the US is the only country that provides nuclear security guarantees to other countries, over 30 nations around the world, providing assurance to American allies—all with the same nuclear arsenal that protects the United States.

Second, currently used categories give researchers wide discretion when it comes to identifying nuclear-related spending. The DoD's budget request includes a budget sub-function, 053 atomic energy defense activities, which incorporates the Department of Energy's defense programs, including the NNSA weapons activities budget.[26] The Office of Management and Budget's FY 2015 most recent estimate for the 053 budget sub-function is $20.837 billion.[27] The 053 budget sub-function also includes defense facilities closure projects, defense environmental restoration and waste management, defense environmental management privatization, other defense activities, defense nuclear waste disposal, and funding for the Defense Nuclear Facilities Safety Board. This general overview illustrates how different assessments of nuclear expenditures can lead to different budget estimates, even if one uses the same budget documents—often inflating the real cost of nuclear weapons-related activities.

Third, the definition of nuclear weapons expenditures is not clear, especially when it comes to warhead modernization. Any nuclear modernization budget assessment is complicated because there is no government "nuclear modernization" function. Rather, different programs are scattered through different agencies, budget functions, sub-functions, programs, and activities. In 2013, Rose Gottemoeller, Under Secretary of State for Arms Control and International Security, stated, "We're not modernizing. We're not modernizing. That is one of the basic, I would say, principles and rules that have really been part of our nuclear view and part of the policy."[28] The 2010 *NPR* says that the US will not build any new nuclear weapons, will not give existing weapons new military missions, and will not develop any new capabilities.[29]

It is a testament to the uniqueness and power of nuclear weapons that Congress and the President continue to provide funding for the enterprise given these constraints. It is hard to imagine Congress supporting research and development of a "new" conventional weapon that is not substantively better than a current capability—an irony often lost on critics of the nuclear arsenal. While the language used in Washington suggests otherwise, arguably the US is not modernizing its nuclear warheads. It is extending the operational lives of current warheads—based on 1970s designs—through LEPs.

Some analysts include costs of indirectly related nuclear weapons activities like environmental cleanup (including in countries other than the US); nonproliferation efforts, including arms control negotiation and treaties implementation; and even costs of defensive measures like missile defense.[30] Aside from domestic environmental cleanup activities, the US would have to pursue many of these and other measures regardless of whether the country is modernizing its nuclear weapons or even possesses nuclear weapons. For example, after signing the Biological Weapons Convention (1972) and ratifying the treaty (1975), the United States terminated its biological weapons program, yet it continued to invest large sums of money on biological weapons research, nonproliferation, defense, and related activities. This was because, much like it will be for a nuclear-free United States, other nations did not eliminate their own biological weapons programs. Even though America took the lead, many—particularly our adversaries—did not follow. Thus, the United States can expect to continue spending many billions of dollars on nuclear weapons, even after the country no longer deploys its own.

Although often done, it is inaccurate to include missile defense costs in nuclear weapons modernization/infrastructure budgets. This often occurs because proponents of such accounting practices view ballistic missile proliferation as a response to US nuclear weapons, which is not the case. Current adversaries seek to exploit American vulnerabilities. A susceptibility to ballistic missile attack, and the ability to use such an attack to deter American action, drives their interest in procuring ballistic missiles. Expectations that others will give up their ballistic missiles are unrealistic in the foreseeable future.

Fourth, critics of the arsenal often point to LEP cost estimates and suggest that these programs are unaffordable—which is incorrect in the context of both the defense and federal budgets. Beyond that, long-term predictions are inherently uncertain. Given the federal government's poor track record of cost forecasting (for both conventional and nuclear programs), using today's data to project costs 30 years into the future is highly inaccurate at best.[31] A part of the problem is the complexity of these systems. Nuclear warheads are comprised of thousands of parts that must work with precision, and the physics surrounding the nuclear package is not completely understood. Often, program cost increases are caused by presidential and congressional decisions to delay programs or their essential parts due to budget pressures that cause delays in accompanying programs.

The B61 LEP is an example of these phenomena. The 2012 cost estimates for the B61 LEP was about $4 billion.[32] The NNSA's May 2013 report put the estimate at $8.1 billion, with $7.3 billion directed to B61 funding.[33] Some of these increases are a result of program delays caused by sequestration and defense budget uncertainty in the coming years, while some of the cost increase is a result of the detailed design and engineering analysis that was not completed for the initial estimate. The DoD's Cost Assessment and Program Evaluation (CAPE) assesses the cost of the program at $10 billion—although this number is in dispute.[34] The difference is caused by diverse assumptions that the agencies use.[35] An additional reason for significant increases in LEP costs is due to the incorporation of a wide array of non-LEP or marginally related basic science activities the weapons labs regularly include in estimates for LEP programs. This practice is seen as an effective means of funding scientific research that would otherwise not receive such funding. As one senior Air Force officer pointed out, this is scientific research with implications for the safety, security, and effectiveness of the nuclear stockpile that would otherwise not receive such funding.[36]

Fifth, some elements of the nuclear triad serve dual purposes. A bomber can be nuclear or conventional. Some facilities in the NNSA complex work on non-nuclear weapons-related projects three-quarters of the time. This raises a question as to what part of their modernization and operation and maintenance budget should be attributed to a nuclear deterrence mission. The problem is most pronounced with the bomber leg of the nuclear triad. Unlike during the Cold War, nuclear-certified bombers today are not on nuclear alert and do not customarily carry nuclear weapons.[37] Nuclear bombs are stored separately from the bombers and a majority of the bombers are dedicated to conventional missions—a condition of the New START Treaty.

Currently, ballistic missile submarines are believed to provide the most survivable leg of the nuclear triad. The *Ohio*-class submarine program is a good example of another budgetary challenge that long-term nuclear weapons modernization estimates face. Today, the US Navy operates 4 of its original 18 *Ohio*-class SSBNs in a conventional mode. Two more SSBNs are scheduled for retirement at decade's end.[38] The future strategic submarine is currently planned with an option for conventional conversions. It would be difficult to estimate what portion of the budget could be attributed to a conventional strategic submarine program because conversion to conventional use is not settled. It is, however, worthwhile to remember that strategic submarines, when converted, can and do serve useful conventional missions. While this is the most expensive leg of the nuclear triad, it costs less than 1 percent of federal expenditures.[39] The SSBN modernization is overdue. As Rear Adm. Joseph Tofalo points out, the

Ohio-class SSBN replacement has been delayed by "more than 20 years from the original expected replacement date."[40]

Command and control (C2) is another problematic category because systems often support both nuclear and conventional forces. Some studies omit these costs altogether. The CBO estimates C2 costs at $56 billion between 2014 and 2023.[41] Since the infrastructure must be reliable and resilient vis-à-vis another nation's actions, it is unlikely that these costs would be significantly different had the US not had its own nuclear weapons arsenal. It is impossible to credibly de-link the C2 nuclear weapons-related costs from the general reliability and resilience requirements based solely on open source information, or the fact that the same system serves the C2 network for conventional operations. This example highlights just how difficult it is to effectively determine the exact cost of the nuclear arsenal.

Sixth, even if a leg of the triad, in full or part, were cut, relative savings would be proportionally smaller because a large percentage of the costs are associated with fixed infrastructure, not just the delivery platforms and warheads that would be cut.[42] And with a number of systems serving a dual-purpose (conventional and nuclear) function, expectation of cost saving resulting from cuts may often go unrealized. Some analysts argue that the B-21 bomber could be unmanned if a nuclear mission was not one of the requirements. It is not clear this would be the case for both technological and cultural reasons. It is also not clear that an unmanned bomber would be cheaper than its manned version considering the technical, technological, and certification obstacles the system would face.

Seventh, when it comes to the follow-on bomber, it will fly conventional missions after its initial deployment and potentially for over a decade after it enters service. The bombers will be nuclear-capable and, within two years of initial operational capability, nuclear-certified—as law requires. Although, some doubt the B-21 bomber will ever be nuclear-certified. Thus, it is questionable to attribute the entire follow-on bomber costs (sometimes including costs of recruiting and retention of the Air Force personnel) to nuclear weapons modernization—as is done in the MIIS $1 trillion dollar 2014–2023 estimate.

The problem of dual-capable systems is even more significant regarding short-range nuclear weapons, which is why most studies omit them. The US currently has an estimated 200–300 B61 gravity bombs in Europe. The costs of forward-deploying them is shared between host countries and the US. This is similar for the costs of dual-capable fighters and personnel trained to maintain and operate them.[43] However, such aircraft are often left out of cost estimates for nuclear weapons expenditures.

It is also worthwhile to keep in mind that while some studies tend to talk about "strategic offensive" systems, US nuclear weapons primarily exist for a deterrence and assurance mission. The requirements for deterrence and assurance can be quite different from requirements for warfighting. Choosing to talk about nuclear weapons as solely "warfighting" weapons misses an important element of US national security. Even more problematic is an inclusion of recruiting and training costs, along with the costs for family housing, since only a small percentage of Navy and Air Force forces remains dedicated to a nuclear mission their entire careers.

Eighth, the administration's own nuclear weapons estimates tend to be lower than estimates listed in the reports discussed above. The MIIS study claims that "the administration and Congress are only now beginning to recognize the full scale of the investments being contemplated and have not yet made a public case for this level of investment."[44] The Obama administration estimated the costs of nuclear infrastructure recapitalization at about $85 billion between FY 2010 and FY 2020.[45] The *Section 1251 Report* went further in its cost estimates of the nuclear enterprise and delivery platform recapitalization, putting the estimate at about $214 billion between FY 2012 and FY 2021.[46] In determining the administration's figures, the White House relied on the 2010 *NPR* which "considered whether the nuclear Triad of SLBMs, ICBMs, and heavy bombers should be retained, and, if so, the necessary investments to sustain each Triad leg."[47] The NPR concluded that the US should retain all three legs of the nuclear triad. It also suggests the triad "will best maintain strategic stability at reasonable cost, while hedging against potential technical problems or vulnerabilities."[48]

James Miller, then-Principal Deputy Under-secretary of Defense for Policy, estimated spending on nuclear weapons (2012–2021) at "$125.8 billion for the delivery systems and about $88 billion for the NNSA-related costs."[49] He criticized higher nuclear weapon estimates as rather imprecise, "suffice it to say there was double counting and some rather curious arithmetic involved."[50] Ashton Carter, then-Deputy Secretary of Defense, was even more vocal in defense of nuclear weapons costs, adding,

> You may all be surprised to know that nuclear weapons don't actually cost that much. Our annual spending for nuclear delivery systems is about $12 billion a year. This is out of around $525 billion, our budget, coming down. And another $4 billion-ish for the command-and-control system that goes with the nuclear weapons, the radar upper, warning, the special communications to make sure that we could—the president could retaliate under any circumstances, especially if we're attacked first, and all that, another $4 billion. So that takes you up to about $16 billion. And so it is not a big swinger of the budget.[51]

According to the MIIS cost estimates, between 2024 and 2029, the US plans on spending about 3 percent of the Pentagon's budget on nuclear delivery vehicle modernization.[52] The report equates this spending to President Reagan's strategic buildup. This is somewhat deceptive since the Pentagon's budget at the time was around 6 percent of GDP while it is a little less than 3 percent today. The DoD under President Reagan spent approximately 39 percent of its budget on modernization. While under the Budget Control Act, the nation is on track to spend about a half that percentage on modernization.[53]

American advocates of nuclear modernization consider cost arguments by critics largely ideological rather than fiscal in nature. As the preceding section suggests, the accounting practices employed are often suspect and designed to maximize arsenal costs. Additionally, the studies highlighted earlier suffer from a lack of context. For example, nuclear weapons-related spending is rarely placed within the larger DoD or federal budget context. This gives a false impression that the US cannot afford its nuclear arsenal and spends too much on it. Efforts to project nuclear weapons spending out three decades, all for the purpose of coming up with a $1 trillion price tag, are unprecedented since there is no accurate way of projecting costs three decades into the future. In short, advocates do not view attacks on the nuclear arsenal based on economic reasons as credible or a contribution to the debate over the role of nuclear weapons in American national security.

Recommendations

It is no surprise that there is limited agreement on federal spending for the larger nuclear weapons enterprise and planned modernization efforts. Fortunately, there are steps the US can take to clarify costs, make spending more transparent, and alleviate growing pressure on the 5 percent of the defense budget dedicated to nuclear weapons programs. The following suggestions offer some initial recommendations.

Congress and the President could achieve perhaps the single greatest cost saving *within* the DoD budget if they modernized the military benefit/entitlement structure. The Budget Control Act, sequestration, and the unwillingness of national leaders to reform ever-increasing benefit/entitlement costs for active duty military members and veterans mean the percentage of the defense budget available for operations, research and development, and related activities is expected to decline in the future. The rising costs of military personnel and DoD civilians (military personnel sub-function) is also crowding out resources for important activities such as procurement and research, development, test, and evaluation. The Stimson Center study notes, "Health care costs and personnel compensation being one of the largest drivers of growing costs, costs which would certainly affect the cost of our strategic nuclear offensive forces."[54] These pressures translate into fewer resources for both nuclear modernization and other defense programs.

Thus, the first order of sound fiscal policy must be a reform of military entitlements—both within the DoD and within federal entitlement programs at large. Absent a comprehensive overhaul of the current inefficient and ineffective benefit/entitlement structure (within the DoD and at large), the US will not have the required resources for nuclear deterrence or defense, regardless of one's opinions on the guns vs. butter debate. While fiscal choices have always played a role in determining defense spending, the US may be driven to assume a defense posture and make defense decisions that are driven by severe resource constraints rather than threats to American interests or the nation's sovereignty. Such a strategic posture should be unacceptable even to critics of nuclear weapons.

Waste, fraud, and abuse within Medicare and Medicaid, two large entitlement programs, averages $65 billion annually—twice the annual cost of nuclear weapons.[55] When it comes to unwisely spent entitlement and welfare dollars, such waste, fraud, and abuse is only the tip of the iceberg. Thus, there is the possibility of finding significant cost savings in these programs, which will easily fund the nation's ultimate insurance policy.

A second reform should tackle the development of a regularized government-wide methodology to assess nuclear weapons expenditures. Currently, the DoD and the NNSA do not have a regularly produced single methodology to assess nuclear weapons costs, although the *Section 1251 Report* does make some progress in this area. This leads to significantly different cost estimates depending of what assumptions are made—leaving open the opportunity to misstate expenditures on nuclear weapons-related programs. A regularly produced single methodology with clearly identified and agreed upon cost estimates would increase transparency of the process. This also seems to be one point where major studies estimating nuclear weapons costs agree.

Congress should disaggregate nuclear weapons modernization expenditures from nuclear sustainment costs—to further increase transparency of nuclear weapons-related expenditures. With the nation's nuclear enterprise no longer focused solely on the nuclear mission, nuclear laboratories advance many civilian missions, including supercomputing and medical research. These efforts, often paid for in part or full with nuclear modernization funding, should not be counted against nuclear weapons programs, even if such funding enables their existence.

Congress should require a comprehensive assessment of systems and nuclear infrastructure that fulfill conventional missions and non-nuclear research tasks but are funded from nuclear modernization accounts. Many within the Beltway intelligentsia, academia, and Congress are unaware of the nation's dual-capability platforms and the non-nuclear purposes of labs within the US nuclear weapons enterprise. This is especially true regarding the bomber force, command-and-control networks, and civilian missions and general science advanced in NNSA-funded labs. A clear delineation would not only help the discussion about the real costs of nuclear weapons but it would help the NNSA spread awareness of its important role in the general advancement of science and technology.

In the end, long-term cost predictions for the nuclear arsenal cannot account for unexpected change in the strategic environment. Such changes are often unexpected and lead to a dramatic shift in requirements and demands placed on American nuclear forces. One of the best examples is the collapse of the Soviet Union which led to an almost 90 percent reduction in American operationally deployed strategic nuclear weapons (from peak Cold War levels). Few predicted such a dramatic change even in 1988. The same can be said of current changes in the strategic environment. Few analysts would have begun 2014 predicting that Russia would invade and annex Crimea all while Vladimir Putin became increasingly aggressive and hostile toward the United States—openly discussing nuclear war.

It is important to maintain a sense of humility and caution when advocating further reductions in the nation's nuclear arsenal because future strategic surprises may be both sudden and unpleasant and may require the US to spend significantly more on its nuclear arsenal in order to maintain peace and stability. And unlike many conventional capabilities, the nuclear force cannot be quickly expanded to meet the demands of a changing strategic environment. This is not to say that long-term budget estimates do not have value. It is important, however, to understand that budget predictions are only a piece of a more complicated national security puzzle, and only a relatively minor one. In the end, the issue is too important to have the outcomes determined by the budget rather than sound considerations of national security.

Notes

1 Barack Obama, Prague Speech on Nuclear Weapons, Prague 2009, www.huffington-post.com/2009/04/05/obama-prague-speech-on-nu_n_183219.html.
2 Stephen Schwartz and Deepti Choubey, "Nuclear Security Spending: Assessing Costs, Examining Priorities," Carnegie Endowment for International Peace, 2009, 6, http://carnegieendowment.org/files/nuclear_security_spending.pdf. All values are in 2014 adjusted dollars.
3 Jon Wolfsthal, Jeffrey Lewis, and Marc Quint, "The Trillion Dollar Nuclear Triad," James Martin Center for Nonproliferation Studies, January 2014, 4, http://cns.miis.edu/opapers/pdfs/140107_trillion_dollar_nuclear_triad.pdf.
4 Russell Rumbaugh and Nathan Cohn, "Resolving Ambiguity: Costing Nuclear Weapons," The Stimson Center, June 2012, 6, www.stimson.org/images/uploads/research-pdfs/RESOLVING_FP_4_no_crop_marks.pdf.
5 Rumbaugh and Cohn, "Resolving Ambiguity," 32.
6 The study does not provide a specific cost estimate for these elements.
7 Rumbaugh and Cohn, "Resolving Ambiguity," 58.
8 Rumbaugh and Cohn, "Resolving Ambiguity," 61.
9 Congressional Budget Office, *Projected Costs of US Nuclear Forces, 2014 to 2023*, December 2013, www.cbo.gov/sites/default/files/cbofiles/attachments/12–19–2013-NuclearForces.pdf.
10 Congressional Budget Office, *Projected Costs of US Nuclear Forces, 2014 to 2023*, 2.
11 The CBO defines "other nuclear-related activities" as legacy costs of nuclear weapons and infrastructure, costs for threat reduction and arms control, and costs for missile defense and other defenses. Congressional Budget Office, *Projected Costs of US Nuclear Forces, 2014 to 2023*, 6.
12 Michaela Dodge, "Why Canada Should Join the US Missile Defense Program: Ballistic Missiles Threaten Both Countries," Heritage Foundation *Backgrounder* No. 2918, June 16, 2014, 9, www.heritage.org/research/reports/2014/06/why-canada-should-join-the-us-missile-defense-program-ballistic-missiles-threaten-both-countries.
13 Dodge, "Why Canada Should Join the US Missile Defense Program: Ballistic Missiles Threaten Both Countries," 18.
14 Wolfsthal, Lewis, and Quint, "The Trillion Dollar Nuclear Triad," 11.
15 General Charles Jacoby, remarks at the Space and Missile Defense Symposium, August 13, 2013, Huntsville, Alabama.
16 The Heritage Foundation, "Where Does All the Money Go?" Federal Budget in Pictures, The Heritage Foundation, 2013, www.heritage.org/federalbudget/where-did-your-tax-dollar-go.

17 US Department of Defense, *United States DoD Fiscal Year 2015 Budget Request Overview*, Office of the Under Secretary of Defense (Comptroller)/Chief Financial Officer, March 2014, http://comptroller.defense.gov/Portals/45/Documents/defbudget/fy2015/fy2015_Budget_Request_Overview_Book.pdf.

18 Jim Martin, *United States Air Force Fiscal Year 2015 Budget Overview* (Washington, DC: Office of the Deputy Assistant Secretary of the Air Force for Management and Budget, 2014), 6; and Stephen W. Wilson, *Air Force Nuclear Enterprise Top Ten* (Shreveport, LA: Air Force Global Strike Command), 23.

19 William K. Lescher, *Department of the Navy FY15 President's Budget* (Washington, DC: Deputy Assistant Secretary of the Navy for Budget, 2014), 5; and Aviation Week and Space Technology, "US Nuclear Triad: You Get What You See," *Aviation Week*, November 11, 2013, http://aviationweek.com/awin/us-nuclear-triad-you-get-what-you-see.

20 US Department of Energy, National Nuclear Security Administration, *FY 2015 Congressional Budget Request*, March 2014, http://energy.gov/sites/prod/files/2014/04/f14/Volume%201%20NNSA.pdf. Currently, US aircraft carriers, three classes of attack submarines, and the Ohio-class strategic submarines use nuclear reactors for their propulsion.

21 We selected the President's fiscal year 2015 budget request because it is the most recent one. Indeed, many budgetary significant actions were taken in the prior years.

22 "Where the Money Comes From … and Where it Goes," *Washington Post*, April 10, 2014, www.washingtonpost.com/wp-srv/special/politics/presidential-budget-2014/.

23 Trent Hamm, "How the Average American Family Spends Their Income—And How to Trim It," The Simple Dollar, August 27, 2014, www.thesimpledollar.com/how-the-average-american-family-spends-their-income-and-how-to-trim-it/.

24 Kenneth N. Waltz, "Why Iran Should Get the Bomb," *Foreign Affairs* (July/August 2012) 91, 4, 2–5.

25 "Nuclear Powers Emerge as US Stockpiles Shrink," The Heritage Foundation, www.heritage.org/~/media/images/reports/2010/solutionsforamerica/vol. 21_750px.ashx.

26 *The Nuclear Matters Handbook*, Office of the Assistant Secretary of Defense for Nuclear, Chemical, and Biological Defense Programs, originally published 1991, www.acq.osd.mil/ncbdp/nm/nm_book_5_11/appendix_I.htm.

27 White House Information Release, "Outlays by Function and Subfunction: 1962–2019," The White House, www.whitehouse.gov/sites/default/files/omb/budget/fy2015/assets/hist03z2.xls.

28 Carnegie Endowment for International Peace, "2013 Carnegie International Nuclear Policy Conference: Morning Plenary Session: Prague 2.0? Deterrence, Disarmament, and Nonproliferation in Obama's Second Term," Transcript, Moderator George Perkovich, April 8, 2013, 21, http://carnegieendowment.org/files/0410carnegie-morning-plenary.pdf.

29 US Department of Defense, *Nuclear Posture Review Report*, April 2010, www.defense.gov/npr/docs/2010%20nuclear%20posture%20review%20report.pdf.

30 Congressional Budget Office, *Projected Costs of US Nuclear Forces, 2014 to 2023*, 1.

31 For example, the F-35 cost estimate nearly doubled between 2007 and 2012. For more information see US Government Accountability Office, "Joint Strike Fighter DoD: Actions Needed to Further Enhance Restructuring and Address Affordability Risks," Report to Congressional Committees, June 2012, http://gao.gov/assets/600/591608.pdf.

32 Dana Priest, "The B61 Bomb: A Case Study in Needs and Costs," *Washington Post Online*, September 16, 2012, www.washingtonpost.com/world/national-security/the-b61-bomb-a-case-study-in-needs-and-costs/2012/09/16/494aff00-f831-11e1-8253-3f495ae70650_story.html.

33 Donald L. Cook, "B61 Life Extension Program and Future Stockpile Strategy," Subcommittee on Strategic Forces, Committee on Armed Services, US House of

Representatives, undated, http://nnsa.energy.gov/sites/default/files/nnsa/2013-10-30% 20HASC-SF%20Cook%20testimony_0.pdf.

34 Kate Brannen, "Pentagon More Than Doubles Cost Estimate for B61 Nuclear Bomb," *Defense News Online*, July 25, 2012, www.defensenews.com/article/20120725/ DEFREG02/307250004/Pentagon-More-Than-Doubles-Cost-Estimate-B61-Nuclear-Bomb. Congressional staff member, interview by Adam Lowther, October 29, 2014, Washington, DC.

35 The NNSA has faced repeated criticism for significant increases in its program estimates, including design flaws that led to additional cost increases in the later building design stages. This suggests that the problem likely is deeper than just different cost assumption methodologies.

36 Air Force General Officer, interview, October 28, 2014.

37 This is why the New Strategic Arms Reduction Treaty counting rules count one bomber as one warhead, even though a B-2 can carry up to 16 nuclear warheads and a B-52 can carry up to 20 nuclear warheads.

38 This means they can carry conventional cruise missiles and special operations forces. For more information, see Elaine M. Grossman, "Future Navy Submarine to Stick With Nuclear Mission," NTI, August 10, 2010, www.nti.org/gsn/article/future-navy-submarine-to-stick-with-nuclear-mission/.

39 Rear Adm. Joseph Tofalo, "The Value of Sea Based Strategic Deterrence," Inside the Navy, *Navy Live*, July 31, 2014, http://navylive.dodlive.mil/2014/07/31/the-value-of-sea-based-strategic-deterrence/.

40 Tofalo, "The Value of Sea Based Strategic Deterrence."

41 Congressional Budget Office, *Projected Costs of US Nuclear Forces: 2014 to 2023*, 2.

42 Congressional Budget Office, *Projected Costs of US Nuclear Forces: 2014 to 2023*, 2.

43 Michaela Dodge, "US Nuclear Weapons in Europe: Critical for Transatlantic Security," Heritage Foundation *Backgrounder* No. 2875, February 18, 2014, www. heritage.org/research/reports/2014/02/us-nuclear-weapons-in-europe-critical-for-transatlantic-security.

44 Wolfsthal, Lewis, and Quint, "The Trillion Dollar Nuclear Triad," 9.

45 News release, "Fact Sheet: An Enduring Commitment to the US Nuclear Deterrent," White House, Office of the Press Secretary, November 17, 2010, www.whitehouse.gov/ the-press-office/2010/11/17/fact-sheet-enduring-commitment-us-nuclear-deterrent.

46 James N. Miller, General C. Robert Kehler, Thomas P. D'Agostino, and Ellen O. Tauscher, "The Current Status and Future Direction for US Nuclear Weapons Policy and Posture," testimony before the Subcommittee on Strategic Forces, Committee on Armed Services, US House of Representatives, November 2, 2011, http://armed services.house.gov/index.cfm/hearings-display?ContentRecord_id=1FE02D2C-4921-4883-89FD-F379DFE70006&ContentType_id=14F995B9-DFA5-407A-9D35-56CC7152A7ED&Group_id=64562e79-731a-4ac6-aab0-7bd8d1b7e890&Month Display=11&YearDisplay=2011.

47 US Department of Defense, *Nuclear Posture Review Report*, 20.

48 US Department of Defense, *Nuclear Posture Review Report*, 21.

49 Miller et al., hearing before the Subcommittee on Strategic Forces.

50 Miller et al., hearing before the Subcommittee on Strategic Forces.

51 Deputy Secretary of Defense Ashton Carter, remarks at the Aspen Security Forum at Aspen, Colorado, July 18, 2013, www.defense.gov/Transcripts/Transcript.aspx? TranscriptID=5277.

52 Wolfsthal, Lewis, and Quint, "The Trillion Dollar Nuclear Triad," 11.

53 Baker Spring, "An Unacceptable Squeeze on Defense Modernization," Heritage Foundation *WebMemo* No. 3417, December 21, 2011, www.heritage.org/research/ reports/2011/12/an-unacceptable-squeeze-on-defense-modernization.

54 Rumbaugh and Cohn, "Resolving Ambiguity," 56.
55 Kathleen King and Kay Daly, *Medicare and Medicaid Fraud Waste and Abuse: Effective Implementation of Recent Laws and Agency Actions Could Help Reduce Improper Payments* (Washington, DC: Government Accountability Office, 2011).

Bibliography

Air Force General Officer, interview, October 28, 2014.

Aviation Week and Space Technology, "US Nuclear Triad: You Get What You See," *Aviation Week*, November 11, 2013, http://aviationweek.com/awin/us-nuclear-triad-you-get-what-you-see.

Brannen, Kate, "Pentagon More Than Doubles Cost Estimate for B61 Nuclear Bomb," *Defense News Online*, July 25, 2012, www.defensenews.com/article/20120725/DEFREG02/307250004/Pentagon-More-Than-Doubles-Cost-Estimate-B61-Nuclear-Bomb.

Carnegie Endowment for International Peace, "2013 Carnegie International Nuclear Policy Conference: Morning Plenary Session: Prague 2.0? Deterrence, Disarmament, and Non-proliferation in Obama's Second Term," Transcript, Moderator George Perkovich, April 8, 2013, http://carnegieendowment.org/files/0410carnegie-morning-plenary.pdf.

Carter, Ashton, Deputy Secretary of Defense, remarks at the Aspen Security Forum at Aspen, Colorado, July 18, 2013, www.defense.gov/Transcripts/Transcript.aspx?TranscriptID=5277.

Congressional Budget Office, *Projected Costs of US Nuclear Forces, 2014 to 2023*, December 2013, www.cbo.gov/sites/default/files/cbofiles/attachments/12-19-2013-NuclearForces.pdf.

Congressional staff member, interview by Adam Lowther, October 29, 2014, Washington, DC.

Cook, Donald L., "B61 Life Extension Program and Future Stockpile Strategy," Subcommittee on Strategic Forces, Committee on Armed Services, US House of Representatives, undated, http://nnsa.energy.gov/sites/default/files/nnsa/2013-10-30%20HASC-SF%20Cook%20testimony_0.pdf.

Dodge, Michaela, "US Nuclear Weapons in Europe: Critical for Transatlantic Security," Heritage Foundation *Backgrounder* No. 2875, February 18, 2014, www.heritage.org/research/reports/2014/02/us-nuclear-weapons-in-europe-critical-for-transatlantic-security.

Dodge, Michaela, "Why Canada Should Join the US Missile Defense Program: Ballistic Missiles Threaten Both Countries," Heritage Foundation *Backgrounder* No. 2918, June 16, 2014, www.heritage.org/research/reports/2014/06/why-canada-should-join-the-us-missile-defense-program-ballistic-missiles-threaten-both-countries.

Grossman, Elaine M. "Future Navy Submarine to Stick With Nuclear Mission," NTI, August 10, 2010, www.nti.org/gsn/article/future-navy-submarine-to-stick-with-nuclear-mission/.

Hamm, Trent, "How the Average American Family Spends Their Income—And How to Trim It," The Simple Dollar, August 27, 2014, www.thesimpledollar.com/how-the-average-american-family-spends-their-income-and-how-to-trim-it/.

The Heritage Foundation, "Where Does All the Money Go?" Federal Budget in Pictures, 2013, www.heritage.org/federalbudget/where-did-your-tax-dollar-go.

Jacoby, General Charles, remarks at the Space and Missile Defense Symposium, August 13, 2013, Huntsville, Alabama.

King, Kathleen, and Daly, Kay, *Medicare and Medicaid Fraud Waste and Abuse: Effective Implementation of Recent Laws and Agency Actions Could Help Reduce Improper Payments* (Washington, DC: Government Accountability Office, 2011).

Lescher, William K., *Department of the Navy FY15 President's Budget* (Washington, DC: Deputy Assistant Secretary of the Navy for Budget, 2014).

Martin, Jim, *United States Air Force Fiscal Year 2015 Budget Overview* (Washington, DC: Office of the Deputy Assistant Secretary of the Air Force for Management and Budget, 2014).

Miller, James N., Kehler, General C. Robert, D'Agostino, Thomas P., and Tauscher, Ellen O., "The Current Status and Future Direction for US Nuclear Weapons Policy and Posture," testimony before the Subcommittee on Strategic Forces, Committee on Armed Services, US House of Representatives, November 2, 2011, http://armed services.house.gov/index.cfm/hearings-display?ContentRecord_id=1FE02D2C-4921-4883-89FD-F379DFE70006&ContentType_id=14F995B9-DFA5-407A-9D35-56CC7152A7ED&Group_id=64562e79-731a-4ac6-aab0-7bd8d1b7e890&Month Display=11&YearDisplay=2011.

News release, "Fact Sheet: An Enduring Commitment to the US Nuclear Deterrent," White House, Office of the Press Secretary, November 17, 2010, www.whitehouse.gov/the-press-office/2010/11/17/fact-sheet-enduring-commitment-us-nuclear-deterrent.

The Nuclear Matters Handbook, Office of the Assistant Secretary of Defense for Nuclear, Chemical, and Biological Defense Programs, originally published 1991, www.acq.osd.mil/ncbdp/nm/nm_book_5_11/appendix_I.htm.

"Nuclear Powers Emerge as US Stockpiles Shrink," The Heritage Foundation, www.heritage.org/~/media/images/reports/2010/solutionsforamerica/vol. 21_750px.ashx.

Obama, Barack, Prague Speech on Nuclear Weapons, Prague 2009, www.huffingtonpost.com/2009/04/05/obama-prague-speech-on-nu_n_183219.html.

Priest, Dana, "The B61 Bomb: A Case Study in Needs and Costs," *Washington Post Online*, September 16, 2012, www.washingtonpost.com/world/national-security/the-b61-bomb-a-case-study-in-needs-and-costs/2012/09/16/494aff00-f831-11e1-8253-3f495ae70650_story.html.

Rumbaugh, Russell, and Cohn, Nathan, "Resolving Ambiguity: Costing Nuclear Weapons," The Stimson Center, June 2012, www.stimson.org/images/uploads/research-pdfs/RESOLVING_FP_4_no_crop_marks.pdf.

Schwartz, Stephen, and Choubey, Deepti, "Nuclear Security Spending: Assessing Costs, Examining Priorities," Carnegie Endowment for International Peace, 2009, http://carnegieendowment.org/files/nuclear_security_spending.pdf. All values are in 2014 adjusted dollars.

Spring, Baker, "An Unacceptable Squeeze on Defense Modernization," Heritage Foundation *WebMemo* No. 3417, December 21, 2011, www.heritage.org/research/reports/2011/12/an-unacceptable-squeeze-on-defense-modernization.

Tofalo, Rear Adm. Joseph, "The Value of Sea Based Strategic Deterrence," Inside the Navy, *Navy Live*, July 31, 2014, http://navylive.dodlive.mil/2014/07/31/the-value-of-sea-based-strategic-deterrence/.

US Department of Defense, *Nuclear Posture Review Report*, April 2010, www.defense.gov/npr/docs/2010%20nuclear%20posture%20review%20report.pdf.

US Department of Defense, *United States DoD Fiscal Year 2015 Budget Request Overview*, Office of the Under Secretary of Defense (Comptroller)/Chief Financial Officer, March 2014, http://comptroller.defense.gov/Portals/45/Documents/defbudget/fy2015/fy2015_Budget_Request_Overview_Book.pdf.

US Department of Energy, National Nuclear Security Administration, *FY 2015 Congressional Budget Request*, March 2014, http://energy.gov/sites/prod/files/2014/04/f14/Volume%201%20NNSA.pdf.

US Government Accountability Office, "Joint Strike Fighter DoD: Actions Needed to Further Enhance Restructuring and Address Affordability Risks," Report to Congressional Committees, June 2012, http://gao.gov/assets/600/591608.pdf.

Waltz, Kenneth N., "Why Iran Should Get the Bomb," *Foreign Affairs* (July/August 2012) 91, 4.

"Where the Money Comes From … and Where it Goes," *Washington Post*, April 10, 2014, www.washingtonpost.com/wp-srv/special/politics/presidential-budget-2014/.

White House Information Release, "Outlays by Function and Subfunction: 1962–2019," The White House, www.whitehouse.gov/sites/default/files/omb/budget/fy2015/assets/hist03z2.xls.

Wilson, Stephen W., *Air Force Nuclear Enterprise Top Ten* (Shreveport, LA: Air Force Global Strike Command).

Wolfsthal, Jon, Lewis, Jeffrey, and Quint, Marc, "The Trillion Dollar Nuclear Triad," James Martin Center for Nonproliferation Studies, January 2014, http://cns.miis.edu/opapers/pdfs/140107_trillion_dollar_nuclear_triad.pdf.

5   The case for a new nuclear weapons arsenal

Eric Y. Moore

Introduction

The US nuclear arsenal was created during a time when the United States had a singular adversary in the Soviet Union and relied upon the concept of mutually assured destruction (MAD) to protect the United States and its allies. MAD relied upon a massive arsenal of nuclear weapons to ensure that Soviet targets could be credibly destroyed in a second strike if a nuclear war were to break out. At the height of the Cold War, the United States maintained over 20,000 nuclear warheads. Today, Russia, China, and other countries still maintain large and expanding arsenals, which present an existential threat to the United States. The very existence of their nuclear stockpiles make America's nuclear triad— bombers, intercontinental ballistic missiles (ICBMs), and sea-launched ballistic missiles (SLBMs)—necessary for continued deterrence. The fall of the Soviet Union not only eliminated the greatest singular threat to the United States, it also created conditions for previously controlled rogue regimes and terrorist organizations to flourish. In a significant change from the days of the Cold War, the United States must now be concerned with the emergence of a nuclear-capable rogue state such as North Korea, the nuclear aspirations of Iran and perhaps Syria, along with the potential of violent non-state actors (VNSAs) to acquire and use nuclear weapons. The United States is not only experiencing a shift in the security environment, but also a shift in domestic support for its nuclear arsenal. Today, it is popular, even chic, to demand the elimination of all nuclear weapons.

Nuclear abolitionists suggest the United States no longer has the will to use nuclear weapons in any case except in response to a large-scale nuclear attack. They also believe that even if the United States were attacked with a nuclear weapon, the US would likely respond with conventional force. There are two reasons for this belief. The first is the ability of the US defense industry to produce technologically advanced non-nuclear weapons capable of destroying almost any target. This reliance on conventional military might is understandable given the United States's recent success in war. Second, abolitionists believe Americans would avoid retaliating with nuclear weapons because of the devastating power. The yields of most warheads are so great that they generate unintentional collateral damage and radioactive fallout. These effects make the current nuclear arsenal "unusable" and ensure "self-deterrence." Arsenal detractors also point to President Obama's April 2009 speech in Prague, Czech Republic, where he said, "The United States will take concrete steps towards a world without nuclear weapons."[1] They also point to former Secretary of Defense Robert Gates's 2010 *Nuclear Posture Review* (NPR) as evidence that the nuclear arsenal is irrelevant in today's security environment. According to the NPR, "The massive nuclear arsenal we inherited from the Cold War era of bipolar military confrontation is poorly suited to address the challenges posed by suicidal terrorists and unfriendly regimes seeking nuclear weapons."[2] Stephen Younger, former director of the Defense Threat Reduction Agency, writes, "Out of concern that any changes in the weapons in our nuclear arsenal would result in a new arms race, the United States continues to maintain an arsenal vastly more powerful than we need."[3]

Ironically, the NPR's challenge to the relevance of the nuclear arsenal also suggests a way ahead. In order to improve the nuclear arsenal's ability to deter and counter current and future security threats, as well as to provide an effective extended deterrent, the United States should develop new nuclear warheads (of variable yields) that augment the current high-yield strategic nuclear inventory. The United States needs to develop and field warheads for ICMBs, SLBMs, and bombers that are maneuverable and low-yield, as well as to resume research and development of nuclear munitions that are capable of deep earth penetration. Stephen Younger supports this argument, "A stockpile in which 90 percent of the weapons had ten kilotons of yield and the remaining 10 percent had five hundred kilotons is compatible with most future targeting requirements."[4]

The development of new nuclear warheads will require American decision makers to abandon their support for "global zero" and overturn the current policy of "no new nuclear weapons." This is necessary to better deter and defeat threats below the existential level and improve the deterrence value of the nation's nuclear arsenal.

Threats

Scott Sagan, a respected scholar, writes, "States exist in an anarchical international system and must therefore rely on self-help to protect their sovereignty and national security."[5] This statement provides some insight into the most significant reason states seek to develop or acquire nuclear weapons—national security. For some autocratic regimes, a significant threat to their national security are the liberal values of the United States. An American penchant for regime change can be described as a double-edged sword. On one edge, the US installs a new regime that is sympathetic to American interests. On the other edge, regime change can lead "rogue" regimes to seek to balance against the overwhelming conventional superiority of the US military. These regimes view nuclear weapons differently than the United States. Rogue regimes view nuclear weapons as an asymmetric counter to US conventional military capabilities.[6]

Three regimes concern the United States the most: North Korea, Iran, and Syria. Recent diplomatic developments with Iran, as well as the uncertain outcome of Syria's ongoing civil war, have reduced, but not eliminated, tensions over their nuclear weapons aspirations and programs. This leaves North Korea as the regime of greatest imminent concern.

North Korea's nuclear weapons program has been a significant worldwide security issue for over two decades. As early as 1992, the United States worked to issue nuclear security guarantees to North Korea in exchange for its compliance with International Atomic Energy Agency safeguards. In 1994 a framework was agreed to provide 500,000 tons of heavy oil and light-water reactors in exchange for North Korea closing its nuclear facilities.[7] This agreement was short-lived. In January 2003, just prior to the US invasion of Iraq, North Korea withdrew from the Nuclear Non-Proliferation Treaty, reopened its nuclear facilities, and reenergized its nuclear weapons program. According to intelligence officials, North Korea detonated its first nuclear weapon in 2006. The detonation produced a low yield and was believed to be a partial failure. However, North Korea has conducted six more tests. After conducting their latest test in February 2013, North Korean officials stated that the device was a miniaturized weapon design.[8] This new design would indicate a significant and unforeseen advancement in North Korea's design capabilities and would highlight intelligence shortcomings concerning North Korea's nuclear program. The ability to gather intelligence within a closed and ethnically homogenous society, such as North Korea, is extremely difficult and heightens the security concern that the North Korean nuclear weapons program poses for the United States.[9]

Along with its nuclear warheads, North Korea has worked tirelessly to develop the Nodong missile. This intermediate-range ballistic missile (IRBM) has a range of 900 miles and is capable of reaching targets in South Korea and Japan, thereby creating an increased security concern for two of the United States's closest allies.[10] North Korea also developed the longer range, three-stage Taepo Dong ballistic missile, which can reach Alaska and Guam and an even longer range Taepo Dong-2 variant is under development.[11] This missile would be North Korea's first ICBM and would be capable of targeting the continental United States.[12] The intelligence community assesses that North Korea possesses approximately ten nuclear weapons, but no operational ICBMs at this time.[13]

The North Korean government has three purposes for its nuclear weapons arsenal. First, the regime seeks to bolster its prestige. Second, North Korea's nuclear weapons allow the regime to engage in coercive diplomacy with more powerful states such as the United States, South Korea, and Japan. Third, North Korea sees its nuclear weapons as an asymmetric counter to the superior conventional military power of the United States—effectively deterring an American attack or invasion.[14]

Although rogue regimes are a threat, intelligence officials suggest that the greatest threat to the United States, allies, and partners is nuclear terrorism.[15] In 2006, Russian Federation President Vladimir Putin and US President George W. Bush jointly announced the creation of the Global Initiative to Combat Nuclear

Terrorism (GICNT) during the G8 summit in St. Petersburg, Russia.[16] The GICNT has 85 participant countries committed to securing nuclear weapon material and to "strengthen global capacity to prevent, detect, and respond to nuclear terrorism through multilateral activities that strengthen the plans, policies, procedures, and interoperability of partner nations."[17]

Terrorist organizations that seek nuclear weapons are unlikely to undertake a large-scale nuclear weapons development program. According to Joan Rohlfing of the Nuclear Threat Initiative, "The shortest path to a terrorist bomb would be for a terrorist organization to steal what it needs to make a nuclear weapon—only a soda can's amount of plutonium or highly enriched uranium (HEU) equivalent to a five-pound bag of sugar."[18] In 1966, three post-doctoral students at Lawrence Livermore National Laboratory (LLNL) participated in an experiment called the Nth Country Experiment. Their task was to design a functioning nuclear weapon without access to any classified information.[19] LLNL physicists determined, "These three physicists, using access to only open source information, were able to design a workable implosion-type weapon in less than three years."[20] They were only missing the necessary nuclear material.

The question remains, where would terrorists get this nuclear material? Rachel Oswald reports that hundreds of nuclear tests were conducted at Semipalatinsk, a former Soviet nuclear test site in Kazakhstan.[21] The Comprehensive Nuclear Test Ban Treaty Organization suggests that 456 atomic and thermonuclear tests were conducted at this site.[22] These tests left behind considerable fissile material. According to Sergey Lukashenko, director of the Institute of Radiation Safety and Ecology at Kazakhstan's National Nuclear Center, despite efforts to clean and secure the site, a "real likelihood exists that more nuclear material remains out there, buried beneath the soil of the Semipalatinsk steppe, unsecured and potentially vulnerable to theft."[23]

The theft of nuclear weapon material from test sites is only part of the concern. Terrorists can steal weapon-grade material from current nuclear powers by overwhelming security measures. In 2013, United States Department of Energy personnel, playing the role of terrorists during a security exercise, successfully stole a substance representing nuclear weapon materiel.[24] Pakistan, with its proximity to Afghanistan and terrorist activity, is especially vulnerable to similar attacks on its nuclear stockpile.[25] There are two especially troubling possibilities in Pakistan. First, the possibility exists that nuclear sites would be attacked by local extremist groups. Second, is the possibility that "radical militants would be able to infiltrate the military or intelligence agencies, giving them a better position to gain access to nuclear materials."[26] Even though Pakistan has taken considerable measures to ensure the security of its nuclear stockpile, threats from terrorists and the consequences of a terrorist with a captured nuclear weapon remain.[27]

Arguments against an improved nuclear weapons arsenal

Despite the consequences that nuclear-armed rogue regimes and terrorists pose to national security, many people remain opposed to the United States's responsible stewardship and potential employment of nuclear weapons. For abolitionists, even the discussion of nuclear weapon benefits, much less discussion of creating new nuclear weapons and capabilities, is seen as threatening. Detractors offer three primary arguments.

First, the development of a new nuclear warhead will be a departure from President Obama's Prague speech (2009) in which he declared a commitment to a world without nuclear weapons, as well as the National Security Strategy (2010) which states, "We are reducing our nuclear arsenal and reliance on nuclear weapons."[28] A new nuclear weapon may also require testing. Nuclear abolitionists highlight the fact that the United States is a signatory to the Comprehensive Nuclear Test Ban Treaty (CTBT), although the Senate has not ratified the treaty. Resumption of nuclear testing would constitute a breach of the treaty and potentially invite other signatories to test nuclear weapons.[29] The United States, the Soviet Union, China, France, the United Kingdom, India, Pakistan, and North Korea have conducted over 2,000 nuclear tests since 1945, but fewer than two dozen since 1992—mostly by India, North Korean, Pakistan, and a few by France.[30] (See Figure 5.1.)

The effects of nuclear testing can be unsettling. Nuclear abolitionists point to a study in the *American Scientist*, which indicates that samples of bone, gland, and other tissue show that specific radionuclides in fallout material create fallout-related cancers.[31]

The second argument offered by abolitionists suggests that a new low-yield nuclear weapon is inherently destabilizing. While a lower-yield nuclear weapon would reduce battle damage, fallout, and collateral damage, these very characteristics would potentially make it more useable and therefore destabilizing. This situation would make it difficult to control war-time escalation and could increase the potential for an adversary to preemptively strike the United States.[32] Abolitionists cite the removal of US battlefield nuclear weapons from Europe as evidence of the inherent destabilizing nature of any weapon with a low yield. Tactical nuclear weapons (also known as non-strategic nuclear weapons) were designed to be employed by American forces on the battlefield, as opposed to strategic nuclear weapons, to defeat "a Soviet conventional attack on Western Europe."[33] US tactical nuclear weapons were removed for several reasons. Because of the weapons' small and portable size, tactical nuclear weapons were susceptible to theft and employment by terrorist organizations. The collapse of the Soviet Union also reduced the need for tactical nuclear weapons in Europe. Russia also used the presence of tactical nuclear weapons as an excuse to avoid further talks on the reduction of its own tactical weapons arsenal. Last, the United States and Europe experienced a growing anti-nuclear movement that argued the United States could maintain its extended deterrence with conventional forces and strategic nuclear weapons. Simultaneously, several North Atlantic Treaty Organization (NATO) members withdrew their support for tactical nuclear weapons due to changes in their domestic politics.[34]

Finally, nuclear abolitionists argue that the president would never use a nuclear weapon, except in a large-scale attack, making the development of new

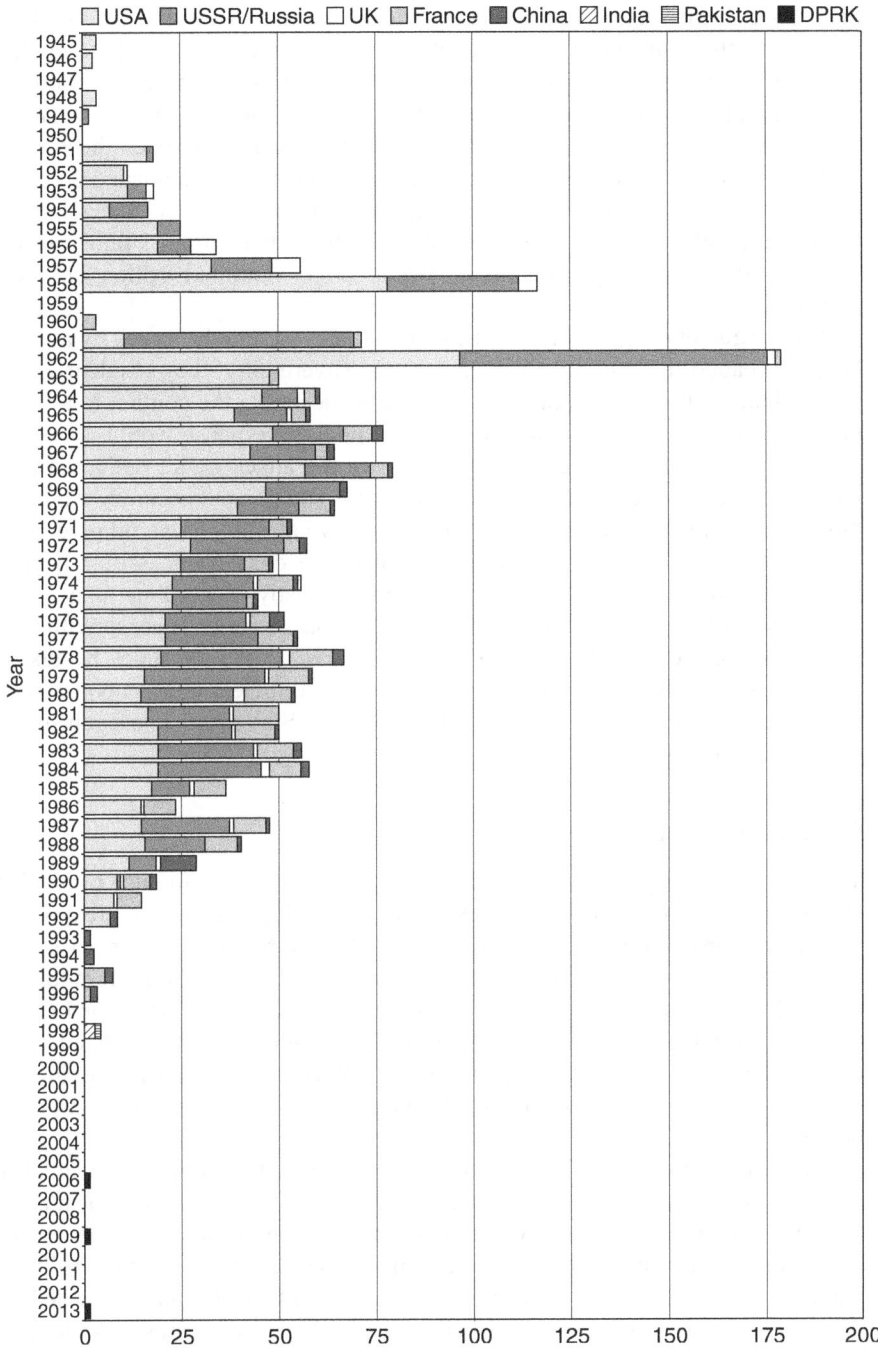

Figure 5.1 Nuclear tests 1945–2013

nuclear weapons and capabilities a waste of resources.[35] This argument suggests that although the United States has significant nuclear capabilities, it lacks the will to use them. In order for the United States to maintain a credible nuclear deterrent it must maintain both the capability to inflict destruction upon the adversary and the will to use that capability. In short, capability plus will equals credibility. One aspect of deterrence without the other neutralizes the nuclear arsenal's deterrent value. For example, suppose North Korea launched an unexpected nuclear attack on the United States. Although the United States has the nuclear capability, would the president have the will to drop a third nuclear weapon on an Asian population or would a conventional option be more palpable?

Their argument suggests that if the United States were to be struck by a single nuclear weapon that did not threaten national sovereignty, the president would not retaliate with nuclear weapons. The fear of escalating the conflict to a larger nuclear conflict is one possible reason why a nuclear response would not come. A second reason is based on a scenario where the attack is not delivered by a ballistic missile. Here, attributing the attack with 100 percent accuracy would be time-consuming. The time it would take to achieve attribution would give political leaders the necessary time to contemplate the consequences of a nuclear response, and be dissuaded from such action. Lastly, according to a senior Department of Defense official, there is a belief that Americans are unwilling to trade New York or Los Angeles for Tokyo or Paris, should the attack be against an ally.[36] It is believed that a nuclear response to a nuclear attack upon an ally would invite a subsequent attack upon the United States and is entirely unacceptable to Americans.

Counter argument

The world has enjoyed relative peace since the development and use of the first atomic weapon. According to Kenneth Waltz, over 60 years without great power war have elapsed since the end of World War II because nuclear weapons elevate the potential costs of great power conflict too high.[37] Even more telling, the total number of casualties (civilian and military) during World War I and II were 96 million people. However, the number of global conflict-related casualties since are down a staggering 89 percent. When one examines the evidence, it is clear that nuclear weapons greatly improve the prospects for great power peace.

Since states seek to maximize their own security, the acquisition of nuclear weapons is an inherently defensive act of "self-help."[38] To ensure that nuclear weapons remain a defensive weapon of "sovereignty insurance," the United States must maintain a capable and credible nuclear arsenal. Nuclear abolitionists have conveniently failed to point out that President Obama acknowledged this reality in his Prague speech when he said, "Make no mistake: As long as these weapons exist, the United States will maintain a safe, secure, and effective arsenal to deter any adversary and guarantee that defense to our allies."[39]

Detractors have also overstated the destabilizing aspects of testing a new nuclear weapon. Not only are advocates of CTBT ratification incorrect in suggesting that below ground tests are destabilizing, but they also seem to forget that India, Pakistan, North Korea, and potentially Iran all became nuclear powers at a time when the established nuclear powers had nearly ceased testing and did not resume testing in response. This calls into question the suggestion that nuclear testing can lead to further proliferation as there has not been an all-out effort by most states to acquire a nuclear weapon.

For the United States, the costs and benefits of nuclear testing present a complex issue. In an off-the-record discussion, a senior National Nuclear Security Administration official stated, "It is possible to use existing designs and detonate only the primary part of the physics package and produce a substantially lower yield while remaining within the guidelines of current designs."[40] Additionally, a new warhead developed using a "gun-type" system similar to that used on the first atomic weapon would not require testing. In fact, designers were so certain of the reliability of such a design that it was never tested prior to use against Japan.[41] The Nth Country Experiment, discussed above, helps to illustrate two points. First, the physics of developing a nuclear weapon were demystified. The post-doctoral students used open-source material to develop a nuclear weapon. Second, a working nuclear weapon can be developed without testing—although testing is useful. In fact, the Israeli government developed a nuclear arsenal without ever testing their weapons.

However, the US may not be able to avoid the eventual need to design, develop, and test nuclear warheads in order to maintain a credible deterrent. It is important to remember that the data currently used in advanced computer modeling are between 25 and 50 years old and were gained at a time when sensors were far inferior to those of today. If the need to test during weapon development is discovered, the US should test without delay.[42] In 2002, the *Washington Post* reported that "the [US] President has not ruled out testing to make sure the stockpile, particularly as it is reduced, is reliable and safe ... but there are no plans to do so."[43]

Despite the nuclear laboratories' reliance on the Science-based Stockpile Stewardship Program to ensure weapon safety and reliability without testing, warhead components are deteriorating with no available substitutes and no data from which accurate models of deterioration can be developed. Nuclear weapon components such as beryllium are now subject to intense environmental and health standards and thus rarely produced. In 2005, the Secretary of Energy's Advisory Board argued that

> dependence on some of these older technologies is starting to burden the Life Extension Program(s) [of current US nuclear weapons], for example, weapons parts that are not acceptable for factory production under modern industrial safety and health regulations or the manufacturers have stopped making their particular product or have gone out of business.[44]

According to a senior Air Force nuclear officer, the United States has taken a 30-year procurement holiday regarding its nuclear inventory.[45] This "holiday" has resulted in a shortage of critical components and those with the knowledge to design and build them. If the United States does not recapitalize its nuclear infrastructure soon, these declining resources will be irretrievable.[46] Designing a new weapon solves many of these issues since it is resourced and built from current materials and energizes both aging and new designers.

Contrary to the concerns of detractors, new nuclear warheads may prevent nuclear proliferation as well as improve stability. A modernized nuclear arsenal enhances extended deterrence by enabling the United States to credibly hold a wide range of targets at risk. By bolstering the capability component of the deterrence equation (credibility = capability + will), allies under the nuclear umbrella have less incentive to develop their own nuclear arsenal.[47] Modernizing the American arsenal with a range of new warheads assures allies of the United States's commitment to extended deterrence.

Low-yield nuclear warheads deployed on ballistic missiles can also increase stability. Russia and the US continue to maintain significant numbers of low-yield or tactical nuclear weapons in their inventory. The Center for Arms Control and Proliferation reports that the United States has approximately 500 tactical nuclear weapons deployed to include 200 or more B-61 bombs within NATO countries.[48] Russia has nearly 2,000 tactical nuclear warheads deployed along the European periphery.[49] A new low-yield nuclear weapon would simply be seen as another warhead, not dissimilar from the ones currently deployed, but would have one fundamental difference. It would be less threatening than current forward-deployed weapons because they would be deployed only on continental United States-based ICBMs and SLBMs instead of continental Europe.

Whether high or low yield, the fact remains that nuclear weapons are in an entirely different category than conventional arms. Leaders of nuclear powers tend to hold them in reserve as weapons of last resort. This explains the absence of a war-time nuclear detonation since 1945 even though there have been several proxy wars and elevated tensions such as the Cuban Missile Crisis. Thus, detractors who suggest that low-yield warheads will become regularized battlefield weapons have failed to learn the lessons of Cold War Europe.

Nuclear abolitionists and minimalists are correct in arguing that there is a lack of political will to use nuclear weapons for any but the most extreme circumstances. President Obama's commitment to a world without nuclear weapons and the National Security Strategy's (2010) statement that the United States will rely less upon nuclear weapons creates security concerns for countries which the United States provides extended nuclear deterrence. The development of a new low-yield nuclear warhead inventory may alleviate some political concerns by demonstrating a commitment to nuclear deterrence and by providing the president more palatable options in a crisis. Today, the president has few options concerning nuclear weapons. He can use the lowest yield option on the B-61. Or he can employ high-yield nuclear weapons and accept any collateral damage and radioactive debris that may result—bringing with it significant political ramifications.

A low-yield *maneuverable* weapon could also reduce the problem of over-flight. Current nuclear weapons deployed on ICBMs are likely to overfly Russia in order to reach a target elsewhere. Overflight increases the risk that Russian leadership will misinterpret a launch as targeted at Russia and possibly result in a Russian nuclear response. However, a weapon that can be launched into a high altitude then maneuver to its target without overflying Russia or other adversarial nuclear-armed countries provides the president with yet another set of viable options.

Recommendations

To achieve a more stable and secure future, American policymakers should abandon their fascination with Global Zero, revive a decaying warhead design program, and commit to the development of new nuclear weapon capabilities. The United States and its allies face new threats from state and non-state actors that present significant challenges for the current nuclear weapons inventory and its ability to achieve effective deterrence. These new challenges can be targeted and frequently deterred with a nuclear arsenal comprised of low-yield maneuverable weapons deployed on ICBMs and SLBMs as well as a low-yield deep earth penetrator.

A maneuverable nuclear warhead has several deterrence benefits. First, ICBMs can now reach targets outside Russia without overflying the country. This lowers the likelihood of misinterpreting a US land-based launch as an attack on Russia. Second, warheads can maneuver around anti-ballistic missile defense sites en route to a target resulting in the need for fewer warheads in order to defeat enemy defenses. Lastly, maneuverable warheads give the president a very flexible nuclear option. It eliminates overflight concerns and retains the ability to reach targets nearly anywhere in the world within 30 minutes.

A modernized nuclear arsenal would be invaluable in deterring offensive action from a rogue state, such as North Korea. If, for example, North Korea gave little warning that it intended to launch a ballistic missile against the United States or an ally, a low-yield maneuverable warhead mated to an ICBM or SLBM would provide the United States with the ability to rapidly reach the target and ensure its destruction without overflying Russia. North Korea also maintains one of the world's most extensive networks of hardened structures and deeply buried facilities.[50] Hardened and deeply buried targets (HDBTs) are extremely difficult to destroy. US forces must have thorough intelligence on the location of HDBTs. Once these facilities are located, aircraft must penetrate hostile and potentially heavily defended airspace to drop guided munitions on the facility. Often, facilities are so deeply buried and hardened that it takes repeated sorties to destroy the facility. When facilities cannot be accurately located and destroyed, US forces attempt to seal off the facility. A low-yield deep earth penetrator, due to its greater blast energy and thermal effects, would be capable of destroying the facility in one sortie and alleviate the requirement for multiple sorties and extremely precise (potentially unattainable) intelligence.

The threat of nuclear terrorism is another difficult challenge for the current arsenal. Terrorists often live among civilian populations. Their proximity to civilians presents a targeting challenge for conventional forces, let alone nuclear forces. However, if terrorists were to acquire a nuclear device or the material to develop a nuclear weapon, they are more likely to be deterred from using a device if the United States were perceived to be more likely to use low-yield warheads.[51] Of course, the most preferred option would be decisive preemptive strikes with long-range conventional munitions and/or US special operations forces to neutralize terrorist leaders and facilities.

The addition of a new low-yield nuclear warhead will improve strategic stability because it enables the United States to better hold at risk non-traditional nuclear targets, while also demonstrating commitment to advanced nuclear capabilities—increasing general credibility. While critics may argue that a low-yield option makes nuclear war more likely, deterrence only works if an adversary believes you have the will to use your arsenal. A new low-yield option will send a clear signal to the nation's adversaries. Scott Sagan writes that "deterrence balances are inherently stable."[52] This philosophy explains the nuclear arms race between the former Soviet Union and the United States in which each super-power sought to balance against the existential threat of the adversary's nuclear arsenal. Low-yield nuclear weapons will not invite an arms race because the weapons are incapable of destroying an adversarial state. The most likely outcome is that the adversary will seek ways to mitigate the effectiveness of this new weapon as all adversaries have done throughout history.

Notes

1 News release, "Remarks by President Barack Obama," White House, April 5, 2009, http://whitehouse.gov/the_press_office/Remarks-By-President-Obama-In-Prague-As-Delivered.
2 Robert Gates, *2010 Nuclear Posture Review*, US Department of Defense, 2010, 11.
3 Stephen M. Younger, *The Bomb: A New History* (New York: Harper Collins, 2009).
4 Younger, *The Bomb: A New History*, 217.
5 Scott D. Sagan, "Why Do States Build Nuclear Weapons?" *International Security* 21, 3 (1996): 57.
6 Office of the Secretary of Defense senior official, interview by author, September 2013, Air War College Nuclear Elective Program.
7 Larry A. Niksch, "North Korea's Nuclear Weapons Program," *Issue Brief for Congress*, 2003, 13.
8 US Intelligence Community Official, interview by author, September 2013, AWC Nuclear Elective Program.
9 US Intelligence Community Official, interview by author.
10 Niksch, "North Korea's Nuclear Weapons Program," 8.
11 Niksch, "North Korea's Nuclear Weapons Program," 8.
12 US Intelligence Community official, interview by author.
13 US Intelligence Community official, interview by author.
14 US Intelligence Community official, interview by author.
15 US Intelligence Community official, interview by author.
16 Nuclear Threat Initiative, "Global Initiative to Combat Nuclear Terrorism," Global Security Newswire, 2013, 1.

17 Nuclear Threat Initiative, "Global Initiative to Combat Nuclear Terrorism."
18 Joan Rohlfing, "Leadership from Congress Critical to Ensuring Global Nuclear Security," Nuclear Threat Initiative, 2013, 2.
19 Michael Levi, *On Nuclear Terrorism* (Cambridge, MA: First Harvard University Press, 2007), 74.
20 Levi, *On Nuclear Terrorism*, 74.
21 Rachel Oswald, "Former Soviet Nuclear Test Site Still Holds Mysteries," Global Security Newswire, 2013, 1.
22 Comprehensive Nuclear Test Ban Organization, "The Soviet Union's Nuclear Testing Program," CTBTO Preparatory Commission, www.ctbto.org/nuclear-testing.
23 Comprehensive Nuclear Test Ban Organization, "The Soviet Union's Nuclear Testing Program."
24 Global Security Newswire, "Mock Terrorists Reach Nuclear Bomb Material in U.S. Facility Drill," 2013, 1.
25 Global Security Newswire, "Pakistan Insists Its Nuke Program Is Fully Secure," 2013, 1.
26 Global Security Newswire, "Pakistan Insists Its Nuke Program Is Fully Secure," 1.
27 Global Security Newswire, "Pakistan Insists Its Nuke Program Is Fully Secure," 1.
28 News release, "Remarks by President Barack Obama"; and Robert Gates, 2010 *Nuclear Posture Review*.
29 US Department of State, Report of the Comprehensive Test Ban Treaty (1996).
30 Comprehensive Nuclear Test Ban Organization, "The Soviet Union's Nuclear Testing Program."
31 Comprehensive Nuclear Test Ban Organization, "The Soviet Union's Nuclear Testing Program."
32 US Office of the Secretary of Defense senior official, interview by author.
33 The Center for Arms Control and Non-Proliferation, "US Tactical Nuclear Weapons in Europe Fact Sheet," www.armscontrolcenter.org/issues/nuclearweapons.
34 The Center for Arms Control and Non-Proliferation, "US Tactical Nuclear Weapons in Europe Fact Sheet."
35 Senior State Department official, interview by author.
36 Senior Air Force nuclear officer, interview by author, September 2013, Air War College Nuclear Elective Program.
37 Scott D. Sagan and Kenneth N. Waltz, *The Spread of Nuclear Weapons* (New York: W.W. Norton and Company, Inc., 2003), 33.
38 Sagan and Waltz, *The Spread of Nuclear Weapons*, 57.
39 News release, "Remarks by President Barack Obama."
40 National Nuclear Security Administration senior official, interview by author, September 2013, Air War College Nuclear Elective Program.
41 A. Fitzpatrick and I. Oelrich, *The Stockpile Stewardship Program: Fifteen Years On* (Washington, DC: Federation of American Scientists, 2007), 10.
42 Fitzpatrick and Oelrich, *The Stockpile Stewardship Program: Fifteen Years On*, 10.
43 Walter Pincus, "New Nuclear Warhead Proposed to Congress," *Washington Post*, 2005, 1.
44 Fitzpatrick and Oelrich, *The Stockpile Stewardship Program: Fifteen Years On*, 59.
45 Senior Air Force nuclear officer, interview by author.
46 Fitzpatrick and Oelrich, *The Stockpile Stewardship Program: Fifteen Years On*, 59.
47 William G. Eldridge, *The Credibility of America's Extended Nuclear Deterrent: The Case of the Republic of Turkey* (Maxwell AFB, AL: Air University Press, 2011), 11.
48 The Center for Arms Control and Non-Proliferation, "US Tactical Nuclear Weapons in Europe Fact Sheet."
49 The Center for Arms Control and Non-Proliferation, "US Tactical Nuclear Weapons in Europe Fact Sheet."
50 US Intelligence Community official, interview by author.

51 Elaine Grossman, "What the U.S. Could Do If Pakistan Loses Control Over Nuclear Weapons," Global Security Newswire, 2.
52 Sagan and Waltz, *The Spread of Nuclear Weapons*, 30.

Bibliography

The Center for Arms Control and Non-Proliferation, "US Tactical Nuclear Weapons in Europe Fact Sheet," www.armscontrolcenter.org/issues/nuclearweapons.

Comprehensive Nuclear Test Ban Organization, "The Soviet Union's Nuclear Testing Program," CTBTO Preparatory Commission, www.ctbto.org/nuclear-testing.

Eldridge, William G., *The Credibility of America's Extended Nuclear Deterrent: The Case of the Republic of Turkey* (Maxwell AFB, AL: Air University Press, 2011).

Fitzpatrick, A., and Oelrich, I., *The Stockpile Stewardship Program: Fifteen Years On* (Washington, DC: Federation of American Scientists, 2007).

Gates, Robert, *2010 Nuclear Posture Review*, US Department of Defense, 2010.

Global Security Newswire, "Mock Terrorists Reach Nuclear Bomb Material in U.S. Facility Drill," 2013.

Global Security Newswire, "Pakistan Insists Its Nuke Program Is Fully Secure," 2013.

Grossman, Elaine, "What the U.S. Could Do If Pakistan Loses Control Over Nuclear Weapons," Global Security Newswire.

Levi, Michael, *On Nuclear Terrorism* (Cambridge, MA: First Harvard University Press, 2007).

National Nuclear Security Administration senior official, interview by author, September 2013, Air War College Nuclear Elective Program.

News release, "Remarks by President Barack Obama," White House, April 5, 2009, http://whitehouse.gov/the_press_office/Remarks-By-President-Obama-In-Prague-As-Delivered.

Niksch, Larry A., "North Korea's Nuclear Weapons Program," *Issue Brief for Congress*, 2003.

Nuclear Threat Initiative, "Global Initiative to Combat Nuclear Terrorism," Global Security Newswire, 2013.

Office of the Secretary of Defense senior official, interview by author, September 2013, Air War College Nuclear Elective Program.

Oswald, Rachel, "Former Soviet Nuclear Test Site Still Holds Mysteries," Global Security Newswire, 2013.

Pincus, Walter, "New Nuclear Warhead Proposed to Congress," *Washington Post*, 2005.

Rohlfing, Joan, "Leadership from Congress Critical to Ensuring Global Nuclear Security," Nuclear Threat Initiative, 2013.

Sagan, Scott D., "Why Do States Build Nuclear Weapons?" *International Security* 21, 3 (1996).

Sagan, Scott D., and Waltz, Kenneth N., *The Spread of Nuclear Weapons* (New York: W.W. Norton and Company, Inc., 2003).

Senior Air Force nuclear officer, interview by author, September 2013, Air War College Nuclear Elective Program.

Senior State Department official, interview by author.

US Department of State, Report of the Comprehensive Test Ban Treaty (1996).

US Intelligence Community Official, interview by author, September 2013, AWC Nuclear Elective Program.

Younger, Stephen M., *The Bomb: A New History* (New York: Harper Collins, 2009).

6 Modernizing the nuclear bomber

A national security imperative

Thomas C. Kirkham

Introduction

The Cold War officially ended over 20 years ago. For many people it meant they were finally free from the fear of nuclear war. The nuclear weapons and delivery systems that long protected them seemed unnecessary. This view is widespread in the United States and is illustrated by the fact that nuclear weapons development programs have fallen into a state of decay since the end of the Cold War. In effect, the United States has taken a "procurement holiday" for the past 20 years and has skipped the designing and fielding of a generation of new nuclear weapons. Instead, the United States has chosen to extend the life span of its nuclear weapons and delivery platforms rather than following 40 years of precedent—designing and building new ones. The same cannot be said for other nuclear weapons states. Nuclear powers such as Russia and China continue to regularly modernize their nuclear arsenals. The emergence of North Korea's nuclear capability and the impending advancement of Iran onto the nuclear scene clearly demonstrate that the Cold War may be over, but nuclear weapons still play an integral role in national security. The United States must recognize that as long as other nations place great value in their nuclear weapons programs, it must do the same.

As the strategic bomber force continues to age it will eventually no longer serve as a credible deterrent. Failure to modernize the nuclear bomber fleet weakens America's long-term deterrent and may even lead to greater nuclear proliferation as allies no longer feel protected by American extended deterrence, leading them to develop nuclear programs of their own. For these and other reasons, strategic bombers remain as relevant today as they were at the height of the Cold War. They continue to play a vital role in the security of the United States and the nation's allies. In short, the US must modernize its strategic bomber force in order to increase the nation's flexible response deterrent and ensure the nation's security. Only then will the country be in a position to address both current and future nuclear threats.

History

Strategic nuclear bombers have played an integral role in the defense of the United States for over sixty years. These long-range aircraft can produce tailored effects on a myriad of targets anywhere on the globe within hours. The need to sustain such a fleet is dictated not only by the role America has assumed as guarantor of global security, but also by geographic reality; the US is separated from its adversaries by vast oceans.[1] Bombers initially entered the scene during World War I. The US steadily developed more advanced bombers over the course of the next eighty years. However, after the collapse of the Soviet Union, the US ceased spending on the development of new long-range strike aircraft. Because enemies like al-Qaeda lacked air forces and air defenses, bomber modernization was not regarded as a high priority. Money that would have been spent during the Cold War to keep long-range strike capabilities robust was spent elsewhere.

Instead, the Air Force elected to upgrade its current fleet of bombers by introducing smart weapons, secure data links, and advanced avionics. However, it has not developed a new bomber for more than 20 years. To underscore the point, in 1960, the US Air Force had 1,515 nuclear-capable bombers in its inventory.[2] Today, the Air Force has only 96 nuclear-capable bombers in service and the average age of the strategic bomber fleet is 33 years old.[3] The fact that the newest B-52s still in service rolled off the assembly line over 50 years ago (1962) is unprecedented in American military history and as threats change, it is not clear that what is left of the heavy bomber force can cope with the military challenges that lie ahead. For instance, China is investing heavily in anti-access and area-denial capabilities aimed at keeping US military forces out of its region. If the US is going to dominate the Pacific, which is quickly becoming central to the new global economy, it must modernize its strategic bomber force.[4]

Attributes of the strategic bomber force

If recent experience is any guide to the future, the timing and locations of international crises will prove extremely difficult to predict. This means that the mobility strategic bombers provide ensures the president has flexible options during a crisis or, if necessary, in time of war. The flexibility provided by bombers far exceeds that of either intercontinental ballistic missiles (ICBMs) or ballistic missile submarines (SSBNs), which make up the other two legs of the nuclear triad.

Strategic bombers can carry a wide variety of weapons from conventional to nuclear and from traditional gravity bombs to long-range standoff weapons like air-launched cruise missiles (ALCMs). Also important is the fact that strategic bombers carry the only variable-yield nuclear weapon, which means that the president can use a low-yield nuclear weapon instead of being constrained by the large yields of intercontinental or submarine-launched ballistic missiles (SLBMs). Additionally, bombers eliminate the need to overfly Russia or China, should the target be elsewhere, and they are the only recallable delivery platform.

To the extent that an attack against an adversary is a function of politics, the military tools employed to support it must be responsive to the president and his need for flexible attack options. Strategic bombers can fly airborne alert, ready to proceed to any target at a moment's notice, or deploy forward as a coercive measure as the president seeks to de-escalate a conflict. Although SSBNs and ICBMs are also responsive, their application in a crisis is very limited and offers the president very few options in an escalation/de-escalation scenario.

Given the bombers' ability to cover great distances quickly, free of the obstructions of surface terrain, the only real challenge they face are anti-aircraft defenses, which are yet to detect the United States's stealth bombers. Because they are mobile and can carry a wide array of weapons, an adversary's ability to plan a defense against American bombers is exceedingly difficult.

Should the United States learn that an adversary has deployed his anti-aircraft forces in the right place to defend against incoming bombers, the mission can be changed, and weapons can be reprogrammed in mid-flight as attacking bombers fly around the threat. The same cannot be said of either ICBMs or SLBMs. They simply do not have the flexibility or responsiveness of the bomber force. Their strengths lie in providing prompt or delayed second-strike capability and surviv-ability, both of which are complementary, not redundant to the bomber leg of the triad.

The final characteristic of the strategic bomber force that makes it the world's best airborne nuclear weapons delivery platform is its ability to signal American intent to adversaries. For deterrence to be effective, it is imperative that a nation be able to send a clear message to the country that is about to be on the receiving end of an American attack or, preferably, paying attention to US messages of deterrence. Nothing demonstrates American resolve better than putting fully loaded strategic bombers on alert or deploying them to a forward base as the spy satellites of a target nation pass overhead. The ability to signal flexibly in a nuclear crisis is a characteristic found only in the bomber force.[5]

By their very nature, the readiness statuses of SSBNs and ICBMs are designed to be stealthy and hidden from view. Consequently, their utility in an escalation/de-escalation scenario is extremely limited.[6] In fact, the range of mis-sions in which either could be employed and the kinds of attacks and weapons effects they could create are very limited. Although initially flushing submarines from port or increasing the alert posture of the ICBM force could signal Ameri-can concern during a crisis, little more can be done with these weapons systems after that to send a clear message to an adversary.

In terms of signaling, strategic bombers also enhance the effectiveness of coercive threats. Absent the ability to clearly communicate both the will and the capability to carry out an attack, coercion does not work.[7] Therefore, to be an effective tool in crisis management, strike assets need to be employable in ways that visibly communicate one's capability, resolve, and restraint. Only nuclear capable bombers can effectively perform this function.

The arguments of critics

According to critics of strategic nuclear bombers, the circumstances in which the United States might employ them are rare and rapidly diminishing. With this in mind and in light of a dwindling Department of Defense (DoD) budget, many argue that the bomber leg of the nuclear triad should be eliminated. In the minds of detractors, bombers are overkill. The costs associated with maintaining nuclear-capable bombers are no longer justifiable.[8] In fact, many critics believe the US needs to focus on a dyad of ICBMs and SSBNs, or a submarine-based monad, to counter current nuclear threats. There are many reasons why detractors believe strategic nuclear bombers are no longer a viable delivery platform, but they can be summarized in two primary categories. First, they argue that nuclear-capable bombers have a destabilizing effect on security. Second, they argue that bombers inherently possess several negative attributes that impede their effectiveness as nuclear deterrence weapon systems.

The first argument—bombers are destabilizing—is based on the fact that they are the least secure leg of the nuclear triad. By their very nature, bombers operate from airfields where nuclear weapons have to be transported from secure weapons storage areas to the flight line for loading. During convoy and upload operations, nuclear weapons are more vulnerable to attack since they are in the open and are no longer protected by hardened storage facilities. In fact, this vulnerability remains until the loaded aircraft becomes airborne. Even then, however, nuclear-loaded aircraft are still vulnerable to accidents due to mechanical failure, extreme weather, or human error. Conversely, ICBMs and SSBNs are much more secure. ICBMs are housed in hardened silos until launch and SSBNs are only vulnerable while in port, otherwise they remain invisible under the protection of deep ocean waters that are distant from population centers.

The second argument offered by detractors is based on the fact that bombers, by virtue of their sheer size and slowness, are vulnerable to enemy air defenses. To combat this vulnerability, the Air Force first outfitted nuclear capable bombers with cruise missiles. Being armed with the ability to launch nuclear weapons outside the range of enemy air defenses certainly made bombers much more viable as a nuclear deterrent, but it also introduced the problem of overflight. Launching missiles over sovereign airspace significantly increases the risk calculus of executing a strike. On one hand, asking permission from a country to overfly their airspace can result in a denial or, if approved, can result in a loss of surprise due to leaked information. On the other hand, if the decision is made to launch without the permission of an affected nation and the missile is either detected by their defenses or goes off course and crashes in their territory, an embarrassing or potentially hostile international incident could occur. Critics point out that if air-launched missile overflight is necessary, then bombers provide little that cannot be accomplished with ICBMs or SLBMs.[9]

The second approach taken by the Air Force to reduce strategic bombers' vulnerability to enemy air defenses was the development of stealth technology. The introduction of the B-2A stealth bomber was a game changer that currently enables the US to penetrate even the most sophisticated enemy air defenses. However, critics argue that stealth technology has escalated tensions between near-peer competitors due to the increase in threat, destabilizing the international security environment. Critics also point out that the high cost of stealth bombers ensured they are no longer a cost-effective nuclear delivery platform—a key argument made by proponents of maintaining the bomber leg of the triad. In fact, critics point out that the Air Force plans to buy 80–100 next-generation stealth bombers at a cost of $550 million each with a total program cost estimated between $40 billion and $60 billion.[10] Even Air Force leaders acknowledge that the high cost of the new bomber could lead to its curtailment or cancellation. Current plans also call for upgrading the warhead on the bomber's ALCM and giving the B-2A stealth bomber the capability to launch it. But, as previously mentioned, critics believe this is a niche that can easily be filled by ICBMs or SSBNs.

Aside from bombers' destabilizing effects, critics also argue that they possess several additional negative attributes. First, bombers provide only a minimal second-strike capability. The nuclear triad was developed during the Eisenhower administration as a result of competition between the Soviet Union and the United States. The primary rationale that drove the development of the nuclear triad was to assure the American public of a second-strike capability in the event of a Soviet preemptive attack. A diversification of delivery platforms would complicate targeting for the Soviet Union, thus ensuring the survival of the nuclear arsenal. However, strategic bombers offer only a minimal second-strike capability and to achieve that, bombers must be placed on alert status so they can be launched prior to inbound missile impact. That, combined with the fact that bombers primarily employ weapons in the kiloton range, not megaton range, means that having the capacity to completely destroy a peer nuclear power is highly unlikely.

Another negative attribute pointed out by detractors is that bombers are highly susceptible to surprise attack while on the ground. Today, nuclear capable bombers are few in number—fewer than 100—and they are stationed at only three bases in the United States. This makes them vulnerable to surprise attack. Critics argue that when bombers are forward deployed to overseas contingency bases this vulnerability is increased even further. They suggest this vulnerability to attack, while on the ground, not only diminishes the deterrent value of bombers, but may even invite attack—another reason to eliminate the bomber leg of the triad.

Furthermore, bomber critics argue that American conventional power makes the case for maintaining the nuclear triad exceptionally dubious. Detractors point out that no adversary of the United States likely possesses the capability to destroy the ballistic missile submarine force and ICBM force, making the bomber leg unnecessary. This argument suggests that nuclear weapons, for all practical purposes, are essentially irrelevant in today's wars, which are against non-state actors and weak states that do not have nuclear arsenals.[11] This makes a nuclear-capable bomber unnecessary.

Critics do recognize that nuclear weapons play a role in war scenarios with other nuclear-armed powers. However, cases where the success of deterrence hinges on the United States's ability to destroy enemy nuclear forces are few and far between. This suggests that the few remaining situations in which the use of nuclear weapons may be required could easily be handled by ICBMs, SSBNs, or conventional forces, and does not justify the need for strategic nuclear bombers.

In the minds of many detractors, it is clear that nuclear weapons have grown less important to American national security due to the nation's overwhelming conventional military superiority. They are also quick to point out that fewer states have revisionist territorial aspirations, much less the capability to act on them. Therefore, nuclear threats are not credible and nuclear weapons unusable in the vast majority of real and imagined military contingencies. In their opinion, these factors explain why spending on nuclear weapons by the Pentagon has fallen from almost 27 percent of the defense budget in 1961 to 4–6 percent today.[12] Justifiably then, since less is asked of nuclear deterrence, it demands fewer delivery platforms—with ICBMs and SSBNs more than adequate to address current and future needs.

Last, critics believe the most basic flaw in building a case for maintaining the triad—to provide a counterforce capability—is that deterrence and extended deterrence can be achieved without it. The theory that extended deterrence requires counterforce capability is a Cold War artifact based largely on misperception of the Soviet threat.[13] They point to evidence that suggests European peace would have held firm without American counterforce threats. Both sides saw a reasonable chance of destruction as too risky for engaging in offensive maneuvers.

Even today, critics argue that neither China nor Russia seems greatly concerned by American counterforce capabilities. They believe that China and Russia view even somewhat vulnerable arsenals as sufficient to deter attack. From their viewpoint, scenarios where counter-value threats fail to deter attacks on allies, but counterforce threats succeed, are becoming difficult to imagine. With this in mind, critics argue that the United States needs to adopt a counter-value targeting strategy thus eliminating the need for maintaining a nuclear triad. In their opinion, maintaining the bomber leg of the nuclear triad in today's threat environment is nothing more than overkill and a waste of taxpayer dollars.

The arguments of supporters

Supporters of the bomber leg argue that that the responsiveness and flexibility of strategic bombers provide a stabilizing effect on international security. They argue that the flexibility of bombers lends capability and credibility to deterrence, the capstone of American foreign policy. Deterrence, which is the prevention from action by fear of consequences, is a state of mind brought about by the existence of a credible threat of unacceptable counteraction.[14] Deterrence strategy is intended to dissuade an adversary from undertaking an action not yet started or to prevent them from something another state desires.[15] In order for deterrence to be effective, a nation not only must have the capability to punish an adversary but must also have the will to carry out an attack. To achieve effective deterrence, the United States must have the capability and, most importantly, the credibility to create the desired effect.[16] The flexibility and responsiveness inherent in strategic bombers exemplify these characteristics making deterrence more likely to achieve the desired effects. Given that there is no margin for error in nuclear deterrence, sending the wrong signal is unacceptable.

While critics assert that deterrence is only effective against other nuclear-armed states and does not apply to conflicts with non-state actors, a case can be made that in fact non-state actors are deterrable. In Figure 6.1, looking at the kinetic-effects pyramid may be the best method for understanding how non-state actors could be deterred.

The highest level of the pyramid is nuclear conflict and the lowest is terrorism. Non-state actors undoubtedly would prefer to operate at the highest possible level but are pushed to the bottom of the pyramid because they lack resources and are weak. The lower an actor is on the pyramid, the harder it is for them to change the status quo. According to Mao Zedong, every terrorist/insurgent aspires to move up the pyramid until he is strong enough to defeat his adversary in a conventional conflict.[17]

According to this logic, it is possible to develop an understanding of non-state actors that attaches rationality to their behavior—the most pressing threat the United States faces today. If an adversary is considered rational, he can be deterred. And, as with states, the success of deterrence depends on determining what the non-state actor values, holding it at risk (capability), and effectively communicating a threat to the non-state actor (credibility).[18] The capability and

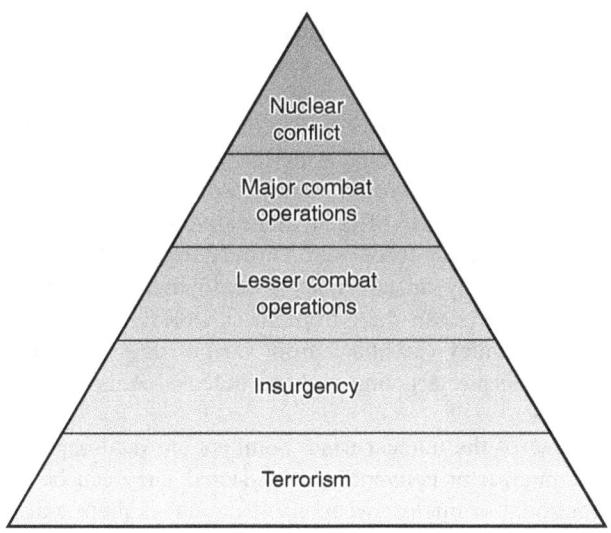

Figure 6.1 Kinetic-effects pyramid

credibility strategic bombers bring to the table make this possible and therefore have a stabilizing effect on international security.

Strategic bombers are also immensely useful in escalation/de-escalation during a conflict.[19] As mentioned previously, ICBMs and SSBNs do not have the ability to demonstrate resolve, another characteristic that has a stabilizing effect on security. Bombers are capable of sending powerful signals to adversaries and allies alike. Governments notice their presence on Guam or other forward deployed locations. They are also keenly aware when they appear over places like South Korea or more recently when B-52s penetrated China's newly proclaimed air defense identification zone located in the East China Sea. Although critics claim that bombers are no longer relevant, events such as these always make front-page news suggesting quite the contrary.

Because bombers are recallable, "scrambling" them toward a potential target is a highly visible way of demonstrating resolve to adversaries and allies without actually launching a nuclear weapon. In a crisis situation, this would enable the cancellation of a bomber strike force after it was ordered, if new information emerged or the president changed his mind.[20] Such a demonstration of resolve might deter a potential adversary and thus prevent war. Land- and sea-based missiles offer no analogous capability. The ability of the US to signal its intentions and resolve singularly hinges upon maintaining the strategic bomber leg of the triad.

Aside from strategic bombers' stabilizing effects on security, they also possess many positive attributes that make them stalwarts of the nuclear triad. Bombers can be dispersed from their bases quickly in order to survive a nuclear first strike.[21] Thus the president would not feel the pressure to "use or lose" bombers during a crisis. Such pressure might not exist with immobile land-based missiles, which are more vulnerable. American bombers can also carry nuclear weapons with the lowest yield, which means that nuclear-capable bombers could potentially provide the president with less devastating options when launching a nuclear attack.[22] This capability would prove extremely valuable if the need arose to strike a non-state actor or engage in a limited strike.

Bombers also minimize the need to overfly Russia and China if the targets were elsewhere. Trying to assure either that American missiles flying towards their country were not meant for them would be difficult at best and would certainly prove destabilizing. Bombers also offer an alternative to missiles in assuring strategic penetration. If a nuclear force were entirely deployed on missiles, that force might be neutralized by an adversary's deployment of a workable missile defense system.[23] Bombers can stage from bases worldwide holding any target on the globe at risk and they can attack from various directions and altitudes. These attributes complicate targeting and the defense plans of potential adversaries.

Finally, like no other leg of the nuclear triad, bombers are dual-capable *and* regularly used for either nuclear or conventional missions. They can be loaded with a myriad of conventional or nuclear weapons. This makes them extremely versatile in conducting a range of military-strike options. This is especially useful in conflict escalation/de-escalation scenarios. Initially, the president can order the use of conventional weapons, however, if the situation escalates and the need arises for a nuclear option, he can increase the deterrence posture of the nation by ordering bombers to be armed with nuclear weapons. The ability of strategic bombers to fill both conventional and nuclear roles not only makes them effective agents of deterrence, it means that the US gets more "bang for the buck," which is crucial in the current fiscal environment.

According to Major General Garrett Harencak, the former Deputy Chief of Staff for Strategic Deterrence and Nuclear Integration, operating costs of the current triad are less than the average American spends at the movies each year.[24] As he has stated, "We're not spending a lot of money on it, and what we are spending is certainly a bargain."[25] This is especially true of the costs associated with maintaining the bomber leg of the triad. Harencak also refuted the notion that the US military is in some sort of Cold War hangover as some critics assert. He stated, "We in no way have anywhere near the infrastructure or even the mindset that we had during the Cold War."[26] The percentage of the defense budget spent on the nuclear enterprise and the number of nuclear weapons and platforms the US possessed during the height of the Cold War far exceeds that of today, making it an affordable and relevant piece of the nation's defense.

Recommendations

With a limited nuclear arsenal at its disposal and considering the critical role the nuclear enterprise plays in effective deterrence, the US must modernize the bomber leg of the triad, while also modernizing the nuclear weapons it employs. The newest platform in the bomber inventory is the B-2, which was initially developed in the late 1980s.[27] B-2s have been in service for over 16 years now and although there are only 19 in the Air Force, they still play an important role in nuclear deterrence. Armed with the B-61, a gravity bomb fielded in 1961, and the newer B-83, the B-2 was designed to deliver weapons over highly defended targets. Despite its somewhat modern technology, the B-2 has two major limitations: the lack of a nuclear standoff capability and its high procurement cost limited the number of aircraft produced.

The B-52 comprises the majority of the bomber fleet. Originally designed in the 1950s, the B-52 is a tribute to its builders and a symbol of decay in the nuclear enterprise. The fact that an aircraft designed almost 60 years ago still plays a major role in the nuclear arsenal demonstrates the neglect that the arsenal has suffered since the end of the Cold War. The Air Force fielded numerous bombers during the Cold War, that were designed to replace the B-52, and yet none has been able to achieve that goal. Even the B-2 has proven to be too expensive to fully replace the B-52 fleet. Instead, the two are now partners in the delicate task of balancing nuclear and conventional missions. Fortunately, the B-52 was so well designed that it still plays a valuable role in the nuclear deterrence mission.

The future of long-range strike (LRS) for the US Air Force focuses on the concept of a B-21 bomber—a long-range bomber to be fielded no sooner than

2018. The NGB is a topic of much interest for the defense industry, the Department of Defense, and Congress. Each has published opinions and considerations pertaining to this future aircraft. There is considerable information pertaining to the necessity of a new bomber. The Congressional Research Service's *Air Force Next Generation Bomber: Background and Issues for Congress* presents a detailed discussion between the Department of Defense and Congress about the necessity of an NGB. The Senate Armed Services Committee found, "Long Range Strike is a critical mission in which the United States needs to retain a credible and dominant capability."[28] Even so, the prospect of building the NGB is at risk due to declining defense budgets.

The B-61 and B-83 gravity bombs employed by the bomber leg of the triad also require modernization. In 1991, the United States built its last nuclear warhead.[29] Since then the nuclear stockpile has been refurbished and modified to stay ahead of problems associated with aging, but no new warheads have come. Stephen Younger, a former senior fellow at Los Alamos, concluded that the effect of these Life Extension Programs (LEPs) on warhead reliability is unknown.[30] Yet a recent unclassified portion of a JASON Defense Advisory Group report seems to agree with the "basic scientific approach" to the LEP.[31] Reports on the reliability of nuclear weapons are classified and are therefore outside the realm of this discussion, but former Secretary of Defense Robert Gates has argued that "the information on which we base our annual certification of the stockpile grows increasingly dated and incomplete."[32] The reality is that nuclear weapons are complicated and it is a tremendous risk to rely on LEPs.

The ALCM, which provides the B-52 with its standoff capability, entered service in the early 1980s and also requires replacement. The US Air Force Life Cycle Management Center Strategic Systems Division (AFLCMC/EBB) is currently conducting market research to identify companies that could participate in the planned long-range standoff (LRSO) cruise missile program. With feedback from industry, the US Air Force is expected to craft an acquisition strategy, which it will then present to the Pentagon for approval. The weapons must be able to penetrate and survive integrated air defense systems from a "significant" standoff range and "prosecute strategic targets in support of the Air Force's global attack capability and strategic deterrence core function," according to the Pentagon's budget justification documents.[33]

The new missile will be designed to be compatible with the B-52H, B-2, and the planned long-range strike bomber (LRS-B). US Air Force budget justification documents show that more than $600 million was requested over the next five years to begin development of the LRSO missile making it another prime target for cancellation due to the current fiscal environment.

The NGB and LRSO programs are vital to effective deterrence and must be fully funded even if it means reducing or eliminating other Air Force acquisition programs. If that were to become necessary, the F-35 program should be considered for reduction. The A-10s and F-16s currently in the inventory are more than capable of filling both close air support and multi-role air-to-ground missions. Additionally, the F-35 program is seven years behind schedule and the cost per aircraft has nearly doubled from $69 million in 2001 to $137 million in 2012.[34]

The reason why the United States must modernize its bomber force is simple. Nuclear weapons and their delivery platforms are needed to deter other nations with nuclear weapons. As long as other nations, allies, or adversaries, maintain and continue to build their nuclear weapons capabilities, the United States must continue to build its own.[35] By not modernizing the bomber fleet and associated weapons, the United States could cause other nations to question the credibility of its nuclear arsenal. For deterrence to have any value there must be a perception, on both sides, that nuclear weapons will be used in certain extreme circumstances and they will function as designed if called upon to do so.

Notes

1 Loren Thompson, "Some Disturbing Facts About America's Dwindling Bomber Force," *Forbes*, August 2013, 1.
2 Natural Resources Defense Council, "Table of US Strategic Offensive Force Loadings."
3 Thompson, "Some Disturbing Facts About America's Dwindling Bomber Force," 2.
4 Thompson, "Some Disturbing Facts About America's Dwindling Bomber Force," 2.
5 Adam Lowther, "Should the United States Maintain the Nuclear Triad?," *Air and Space Power Journal* (Summer 2010): 27.
6 Adam Lowther, "Should the United States Maintain the Nuclear Triad?" 27.
7 Robert Jervis, Richard N. Lebow, and Janice G. Stein, *The Psychology of Deterrence* (Baltimore: Johns Hopkins University Press, 1985), 128.
8 Benjamin Friedman, Christopher Preble, and Matt Fay, "The End of Overkill? Reassessing U.S. Nuclear Weapons Policy," CATO Institute, 2013, 2.
9 Friedman, Preble, and Fay, "The End of Overkill? Reassessing U.S. Nuclear Weapons Policy," 11.
10 Friedman, Preble, and Fay, "The End of Overkill? Reassessing U.S. Nuclear Weapons Policy," 17.
11 Friedman, Preble, and Fay, "The End of Overkill? Reassessing U.S. Nuclear Weapons Policy," 10.
12 Friedman, Preble, and Fay, "The End of Overkill? Reassessing U.S. Nuclear Weapons Policy," 13.
13 Friedman, Preble, and Fay, "The End of Overkill? Reassessing U.S. Nuclear Weapons Policy," 12.
14 US Department of Defense, Joint Publication 1–02, Department of Defense Dictionary of Military and Associated Terms 2010, 73.
15 "Deterrence Theory," Wikipedia, http://en.wikipedia.org/wiki/Deterrence_theory.
16 Adam Lowther, *Understanding Deterrence: Essential Questions for the Twenty-First Century* (Maxwell AFB, AL: Air Force Research Institute, 2009).
17 Lowther, *Understanding Deterrence*, 5.
18 Lowther, *Understanding Deterrence*, 5.
19 Lowther, "Should the United States Maintain the Nuclear Triad?" 27.
20 Kingston Reif and Travis Sharp, "Pruning the Nuclear Triad? Pros and Cons of Submarines, Bombers, and Missiles," Arms Control Center, May 2013, 3.
21 Reif and Sharp, "Pruning the Nuclear Triad?" 6.
22 Reif and Sharp, "Pruning the Nuclear Triad?" 6.
23 Reif and Sharp, "Pruning the Nuclear Triad?" 6.
24 Mark Thompson, "Triad and True…," *Time*, June 2013, 2.

25 Thompson, "Triad and True…," 2.
26 Thompson, "Triad and True…," 2.
27 US Air Force, *United States Air Force Fact Sheet*, www.af.mil/information/factsheets/index.asp.
28 Ronald O'Rourke, "Air Force Next Generation Bomber: Background and Issues for Congress," Congressional Research Service, October 2009, 23.
29 Office of the Deputy Assistant to the Secretary of Defense for Nuclear Matters, *US Nuclear Deterrence*, www.acq.osd.mil/ncbdp/nm/USNuclearDeterrence.html.
30 Stephen Younger, *The Bomb* (New York: Harper Collins Publishers, 2009), 97.
31 National Nuclear Security Administration, Public Affairs, "NNSA Thanks JASONs for Technical Review of LEP Programs," http://nnsa.energy.gov/news/print/1272.htm.
32 Robert Gates, "Gates: Nuclear Weapons and Deterrence in the 21st Century," *Carnegie Endowment for International Peace*, Federal News Service, October 2008, 6.
33 Doug Richardson, "USAF to Move Ahead with Long Range Standoff (LRSO) Cruise Missile," *IHS Jane's 360*, November 2013, 1, www.janes.com/article/29481.
34 Michael J. Sullivan, "F-35 Joint Strike Fighter: Restructuring Has Improved the Program, but Affordability Challenges and Other Risks Remain," *GAO Testimony Report*, June 2013, 13.
35 American Association for the Advancement of Science, "Nuclear Weapons in 21st Century US National Security," Office of the Deputy Assistant to the Secretary of Defense for Nuclear Matters, December 2008, http://aps.org/policy/reports/popa-reports/upload/nuclear-weapons.pdf.

Bibliography

American Association for the Advancement of Science, "Nuclear Weapons in 21st Century US National Security," Office of the Deputy Assistant to the Secretary of Defense for Nuclear Matters, December 2008, http://aps.org/policy/reports/popa-reports/upload/nuclear-weapons.pdf.

"Deterrence Theory," Wikipedia, http://en.wikipedia.org/wiki/Deterrence_theory.

Friedman, Benjamin, Preble, Christopher, and Fay, Matt, "The End of Overkill? Reassessing U.S. Nuclear Weapons Policy," CATO Institute, 2013.

Gates, Robert, "Gates: Nuclear Weapons and Deterrence in the 21st Century," *Carnegie Endowment for International Peace*, Federal News Service, October 2008.

Jervis, Robert, Lebow, Richard N., and Stein, Janice G., *The Psychology of Deterrence* (Baltimore: Johns Hopkins University Press, 1985).

Lowther, Adam, *Understanding Deterrence: Essential Questions for the Twenty-First Century* (Maxwell AFB, AL: Air Force Research Institute, 2009).

Lowther, Adam, "Should the United States Maintain the Nuclear Triad?" *Air and Space Power Journal* (Summer 2010).

National Nuclear Security Administration, Public Affairs, "NNSA Thanks JASONs for Technical Review of LEP Programs," http://nnsa.energy.gov/news/print/1272.htm.

Natural Resources Defense Council, "Table of US Strategic Offensive Force Loadings."

Office of the Deputy Assistant to the Secretary of Defense for Nuclear Matters, *US Nuclear Deterrence*, www.acq.osd.mil/ncbdp/nm/USNuclearDeterrence.html.

O'Rourke, Ronald, "Air Force Next Generation Bomber: Background and Issues for Congress," Congressional Research Service, October 2009.

Reif, Kingston, and Sharp, Travis, "Pruning the Nuclear Triad? Pros and Cons of Submarines, Bombers, and Missiles," Arms Control Center, May 2013.

Richardson, Doug, "USAF to Move Ahead with Long Range Standoff (LRSO) Cruise Missile," *IHS Jane's 360*, November 2013, www.janes.com/article/29481.

Sullivan, Michael J., "F-35 Joint Strike Fighter: Restructuring has Improved the Program, but Affordability Challenges and Other Risks Remain," *GAO Testimony Report*, June 2013.

Thompson, Loren, "Some Disturbing Facts About America's Dwindling Bomber Force," *Forbes*, August 2013.

Thompson, Mark, "Triad and True...," *Time*, June 2013.

US Air Force, *United States Air Force Fact Sheet*, www.af.mil/information/factsheets/index.asp.

US Department of Defense, Joint Publication 1–02, Department of Defense Dictionary of Military and Associated Terms 2010.

Younger, Stephen, *The Bomb* (New York: Harper Collins Publishers, 2009).

7   ICBMs

Cold War relics or products for peace?

Anita Feugate-Opperman

Introduction

The United States initially developed its land-based intercontinental ballistic missile (ICBM) fleet to increase the credibility of nuclear deterrence during the Cold War. Since the Cold War ended the threat has changed, but persists. Critics, however, contend ICBMs are no longer relevant in the post-Cold War world and, in a fiscally constrained environment, the cost of modernizing the system outweighs any security benefit it provides; thus, retirement of ICBMs is warranted. Contrary to this view, ICBMs continue to serve as a key contributor to national security and must remain in the arsenal to deter against both current and future threats.

Background

The United States Air Force (USAF), through Air Force Global Strike Command (AFGSC), is responsible for operating, maintaining, and securing the nation's ICBM force. ICBMs became part of the US nuclear arsenal in the 1950s with the deployment of the Titan and Atlas missile systems followed by the first Minuteman system in the early 1960s.[1] The Minuteman III, deployed in 1970, is projected to remain in service until at least 2030.[2] At one time, there were 1,054 ICBMs on alert.[3] Today, the ICBM force consists of 450 Minuteman III on-alert missiles.[4] In order to comply with New START treaty requirements, the Department of Defense plans to reduce the number of deployed (on alert) Minuteman III missiles to 400 and place the remaining 50 into non-deployed status by removing the ICBMs from the launch facilities.[5]

ICBMs are located in hardened launch facilities designed to protect against nuclear attack. The launch facilities are dispersed across a total of 31,900 square miles in Colorado, Montana, Nebraska, North Dakota, and Wyoming.[6] Each missile and associated launcher, or silo, is connected to an underground launch control center through a system of hardened and pressurized cables. Launch crews, consisting of two officers, perform around-the-clock alert in the launch control centers and are ready to respond to a presidential emergency war order launch at a moment's notice.[7]

ICBMs are Cold War relics

Because the strategic environment has changed dramatically since the Cold War, critics contend ICBMs are no longer relevant and should be retired. The most vocal ICBM critics suggest the weapon has a destabilizing effect on international security, the primary reason it should be removed from the nuclear arsenal. Detractors suggest the weapons are on hair-trigger alert, which is not only desta-bilizing, but also an accident waiting to happen.[8] They go on to claim that a rapid-response posture allows for the possibility that the United States could launch a nuclear attack based on incomplete information or misinterpretation of data, which is a far greater risk than an "out of the blue" nuclear attack.[9] The Global Zero Commission, a strident critic of the ICBM, writes, "Given the end of the Cold War, it makes sense to end the Cold War practice of preparing to fight a large-scale nuclear war on a moment's notice."[10] The widely held view is that Russia, America's only nuclear peer, is no longer a threat so there is no longer a need to respond rapidly. If there is a nuclear threat, submarine-launched ballistic missiles (SLBMs) and/or bombers will have enough advanced notice to successfully respond.

Another destabilization argument suggests that the existence of ICBMs could force adversaries to attack the US mainland to prevent an American retaliatory strike. Critics charge that ICBMs are a Russian "warhead sponge."[11] Because Moscow, or Beijing, will need to attack American territory to eliminate a large part of the nuclear arsenal, Washington will respond in kind and attack Russia's cities, moving toward nuclear annihilation.[12]

Finally, since some ICBMs may not survive a first strike, the president may decide to launch on the warning of an attack instead of waiting to verify an actual attack is occurring.[13] The "use it or lose it" attribute of the ICBM leads to the possibility America may launch a strike before ensuring there is truly a threat. ICBM detractors argue that not only do ICBMs promote greater instab-ility, but other characteristics also make them no longer relevant.

ICBMs were developed as a deterrent against the Soviet Union. Therefore, their optimal post-launch flight path sends them north over the pole and then, in most scenarios, over Russia.[14] If Russia is not the target of a nuclear strike, con-ventional wisdom suggests the United States would most likely not risk Russia misinterpreting an ICBM overflight as an attack on Russia, in an attack against North Korea or Iran. This gives ICBMs little utility against rogue nations if the United States constrains its employment of the weapon to ensure Russia is not inadvertently provoked.

In addition to overflight concerns, detractors argue that ICBMs are not rel-evant because they cannot de-escalate a nuclear crisis. Due to the size of the weapon and its effects, the ICBM has little utility in a limited nuclear conflict. Again, working from the point of view that the United States's most likely adversaries are not peer competitors, the most logical application of nuclear weapons would be several small weapons (most likely delivered from a nuclear bomber) used, as Kenneth Waltz writes, "to produce sobriety in the leaders of all of the countries involved and thus bring rapid de-escalation."[15] Of even more concern for ICBM opponents, once the missile is launched there is no way to either retarget it if the location of a threat has changed, or, more importantly, recall the missile if the adversary responds to the United States's threat and a nuclear attack is no longer justified.[16] Using ICBMs could also cause a bad situation to become worse. Critics charge, the United States cannot use ICBMs effectively to stabilize a crisis on the verge of erupting or visually assure allies they are covered by the nuclear umbrella.

A separate criticism of the ICBM force relates to the expense of modernizing the aging fleet. The Minuteman III has been in continuous operations for over 40 years, but requires a $5 billion life extension to keep it operational through 2030.[17] An additional obstacle facing the ICBM fleet is the fact that many of the manufacturing processes and techniques used to build Minuteman III missile components are no longer available, since many of the companies that originally built components are no longer in business.[18] The United States chose to halt testing of nuclear-specific components following President Clinton's signing of the Comprehensive Test Ban Treaty even though the Senate has not ratified the treaty.[19]

ICBM maintenance and modernization costs in conjunction with the United States's budget concerns helps bolster detractors' argument that ICBMs are no longer relevant for national and international security. In addition, some would argue ICBMs are too inflexible, since they cannot serve any role other than that of a nuclear deterrent.[20] The argument suggests there is no room for a weapon system that does not have both a nuclear and conventional function.[21] On the other hand, US planning for the availability of conventional prompt global strike (CPGS) options could include the use of ICBMs in this mission.

These arguments against the ICBM's relevancy in the current global security environment appear logical and give the impression that the world would become a safer place without them. However, if America were to eliminate the ICBM leg of the triad, the world would be less, not more, secure. In fact, the threat of attack would be greater and the costs of an alternative (conventional or nuclear) would prove more taxing on the American treasury.

ICBMs: a product for peace

The 2012 Defense Strategic Guidance states that "we [the United States] will field nuclear forces that can under any circumstances confront an adversary with the prospect of unacceptable damage both to deter potential adversaries and to assure US allies that they can count on America's security commitments."[22] ICBMs ensure stability and balance in national and global security because they guarantee an adversary cannot deliver a preemptive strike without depleting their arsenal, while the United States remains capable of delivering a retaliatory strike.[23]

Before turning to the specific attributes which make the ICBM such a stabilizing force, one must understand why the ICBM is a relevant weapon system

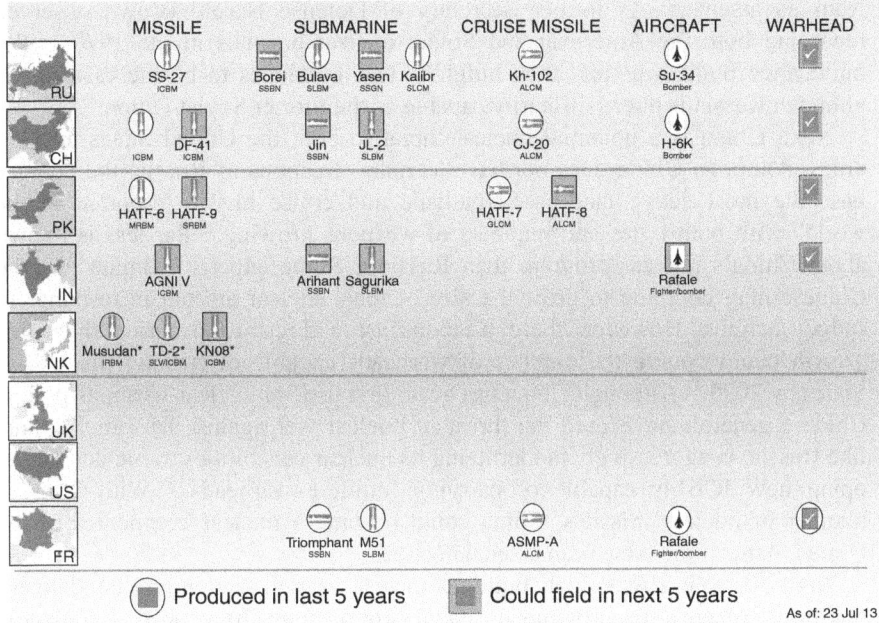

Figure 7.1 Status of nuclear countries' weapons and delivery platforms modernization[24]

across the spectrum of threats in the post-Cold War strategic environment. The United States, with its unique global responsibilities, needs to be prepared to counter traditional threats (Russia), emerging threats (China), rogue nations, and nuclear aspirants. Of note, as illustrated below, the United States and United Kingdom are the only nuclear powers not modernizing or developing new nuclear weapons or delivery platforms.

First, although America has collaborated with Russia on reducing the size of each country's nuclear arsenal, Russia still views a nuclear arsenal, specifically ICBMs, as central to national security.[25] This is in part because of its inferior conventional force, which requires Russia to rely on its long-range and other nuclear forces to defend the nation.[26] Russia's nuclear doctrine has a clear first use of nuclear weapons policy to protect the security of the country even if engaged in conventional warfare under certain circumstances involving especially grave threats to Russia.[27] Moscow also views nuclear weapons as a "guarantee of independent foreign policy and a means of deterrence of ideological competitors."[28] Nuclear weapons continue to be the cornerstone of their warfighting strategy. Russia is not afraid to use threats of a nuclear attack to pressure her neighbors to comply with demands, such as keeping US missile defense systems out of Poland or preventing Ukraine from joining NATO.[29]

Furthermore, Russia has been actively working to modernize its ICBM infrastructure, including a new heavy ICBM capable of carrying up to 15 warheads each; clearly signaling the Russian government does not plan to remove ICBMs from its arsenal.[30] As former Secretary of Defense Harold Brown observed regarding both the American and Soviet nuclear arsenals in the 1970s, "We build, they build, we stop, they build."[31] This continues to be the case today, although Russia is not as militarily capable as the former Soviet Union.[32]

Next, China is a potential nuclear threat to both the United States and her allies. Although China does not have as many weapons as Russia, the country has "the most active land-based ballistic and cruise missile program in the world" with both types and numbers of weapons growing.[33] Far less is known about China's nuclear program than Russia's. Some experts estimate that the Chinese may continue to grow the size of their nuclear arsenal in response to India's actions. However, these weapons are a threat to America since this growth could double the number of warheads capable of striking the United States by 2025.[34] Although China has a "no first use" of nuclear weapons policy, Chinese generals have used the threat of nuclear war against Taiwan.[35] China, like Russia, is aggressively modernizing its nuclear capabilities to include developing new ICBMs capable of carrying multiple warheads.[36] With a rising number of nuclear missiles, China could become a nuclear competitor of the United States if the American arsenal declines.

North Korea is also actively building nuclear weapons and improving delivery platforms, placing the program on equal footing with the country's economic development.[37] Although it may be years before North Korea has proven capability to deliver nuclear weapons, North Korea's Supreme Leader, Kim Jong-un, is already threatening America with a "massive nuclear disaster" and "all out war" if war breaks out on the Korean peninsula.[38] North Korea's hubristic rhetoric notwithstanding, it poses the threat not only of deliberate conventional or nuclear attack, but also of a potential regime disintegration and post-regime chaos with uncertain control over nuclear weapons.

Iran, another foe, does not currently have nuclear weapons, but has voiced its interests in pursuing nuclear weapons and could have the capacity in the near future to develop a weapon.[39] It is thought that Syria was working toward developing its own nuclear capability, possibly with the help of North Korea, but it currently does not have the industrial capacity to pursue developing its own weapons.[40] With the current state of turmoil in Syria, it would be hard to determine if the regime feels threatened enough to reconsider pursuing nuclear weapons. Even though these three countries are not currently existential threats, the possibility of that changing is real. Conventional threats will not always deter rogue nations. There may come a time when a nuclear threat against an enemy regime is necessary. This range of strategic challenges makes the ICBM force relevant for the long term.

Detractors arguing that this rapid response is a large reason ICBMs are destabilizing. This is incorrect. Suggesting ICBMs are sitting on hair-trigger alert is intentionally misleading, and serves only to excite the public by giving the impression that at any moment an ICBM can be launched without permission. While the ICBM force is always on alert and ready to respond promptly, there are multiple layers of controls ensuring launch could only occur upon presidential direction.[41] And, notwithstanding some issues of personnel reliability investigated by the US Air Force in 2014, the Minuteman III has a superb safety record with no incidents that have had the potential to allow for accidental or unauthorized launch.[42]

There are two primary reasons ICBMs must remain on constant alert. First, although the relationship between Russia and the United States is much improved since the Cold War, the two countries still have differences. Russia maintains the alert capability to launch an attack against the United States or her allies. Although many may believe Moscow does not have malicious intent that extends to a nuclear attack against the United States or its allies, this can change, so the US must be ready to respond.[43] Additionally, America has assured her allies that she will come to their aid in the event of a nuclear attack. ICBMs on alert are a continual reminder to allies that America stands by its pledge.[44] Without this assurance, allies may feel the need to develop their own nuclear capability or shift their alliance to another nuclear country, potentially an American nuclear foe, to serve as "nuclear guarantor."[45]

ICBM opponents argue that strategic land-based missiles are destabilizing because ICBMs force adversaries to target the United States homeland. This critique is flawed. While detractors are accurate in saying that they cause adversaries to target the US, they are mistaken in suggesting this is destabilizing. In reality, the fact that an adversary must target the United States is a tremendous deterrent against the use of nuclear weapons because an adversary will need to think long and hard about the consequences of striking another country's sovereign territory.[46] Potential adversaries need only look at how the United States responded to the September 11, 2001, and Pearl Harbor attacks to get an idea of how Americans would react if attacked by a nuclear weapon.[47] American satellites are also able to detect the location of an ICBM launch and thereby easily attribute the attack to the responsible country, leaving no doubt that Washington would know who is responsible.[48] Targeting American ICBM fields also forces adversaries to think twice before launching an attack and by doing so it compels them to realize the benefits of attacking the United States are outweighed by the costs.[49]

Finally, the sheer number of ICBMs and the size of their dispersed locations present a targeting problem for Russia or any adversary. Without ICBMs, adversaries only need to target as few as 13 locations to render America's nuclear arsenal useless.[50] SLBMs have one base on each coast and the bomber fleet is in three locations, so it is easy to see how a nuclear power could readily threaten the non-ICBM nuclear force conventionally. The figure below shows the number of warheads required to eliminate the US ICBM force and non-ICBM targets, illustrating that without ICBMs, just about any nuclear nation could be America's nuclear peer.[51]

As previously stated, American ICBMs are located over a large area making it improbable that a nuclear peer could destroy the entire force in one salvo with any hope of having a meaningful arsenal left for contingencies, reprisal, or

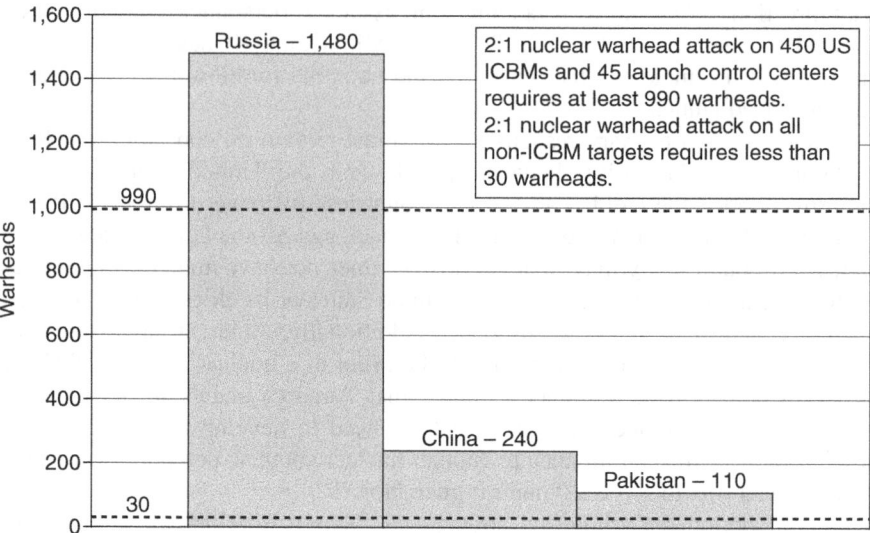

Figure 7.2 2:1 attack on ICBM vs. non-ICBM targets

blackmail.[52] This, in effect, counters the argument that with ICBMs the United States may be inclined to launch on warning instead of actual attack, because a significant fraction of the ICBM fleet will likely survive an attack. And, in the event that competitive Russian or other offensive force building appears to threaten future Minuteman survivability, the Minuteman force can be expanded and, if necessary, defended with affordable and available missile defense technologies.

Each Minuteman III can hold up to three warheads, multiple independently targetable reentry vehicles (MIRVs), but as an additional hedge to global nuclear stability the United States is placing a single nuclear weapon on each missile.[53] The theory of "de-MIRVing" is to make ICBMs a less lucrative target. Since the launch facilities are hardened structures, an adversary will need to expend at least two warheads to destroy a single US warhead.[54] In order to destroy the US ICBM forces, Russia would need 900–1,350 weapons; with a New START treaty limit of 1,550 total warheads, a debilitating attack on the United States would leave Russia with little in reserve.[55] Instead, it is far better for an adversary to focus on destroying a US second-strike capability because allowing it to remain intact would ensure total destruction—a reason to give any adversary pause when contemplating an attack against the United States.

One factor to look at when determining ICBM relevancy is reliability. ICBMs performed well in the past and can be counted on to work as expected if launched against as adversary.[56] The Minuteman III ICBMs have an alert rate of nearly 100 percent, meaning nearly all of its fleet is configured and available for launch at any time.[57] Although it is not possible to test launch an ICBM from operational missile fields, the ICBM community does extensive testing of each portion of the system to ensure it will work if launched. Several times a year, there are test launches of non-nuclear-tipped ICBMs from Vandenberg Air Force Base in California. US Strategic Command also conducts command-and-control exercises to ensure presidential directions are properly sent to the nuclear force and responded to appropriately.[58] Although ICBMs cannot serve as a visual signal of American resolve, like bombers, reliability is a signal to both allies and adversaries that the United States is always ready to respond to a nuclear threat. In fact, a country such as Japan may prefer allowing the US to provide extended deterrence via a reliable, non-visible ICBM force to avoid having an observable US nuclear presence in their country.[59]

Another key attribute of ICBMs is the ability of missile combat crews to quickly retarget their missiles if threats change or an unforeseen target emerges. The system was designed in part to allow the missile combat crew to quickly retarget their missiles and then promptly launch them. Therefore, the operating system and crew reaction allows decision makers time to determine exactly whether a nuclear attack is needed and what it should target.[60] Given the Minuteman III's accuracy—circular error probable (CEP)[61]—the president can use either a single or a small number of ICBMs launched at a precise location, localizing effects and signaling US intent to keep the conflict smaller with a goal of de-escalating the crisis.[62] At the same time, this would show American resolve and capability to continue up the escalatory ladder as needed.

Finally, detractors contend that overflying Russia or China would make using an ICBM a poor choice for the president. However, if the situation required an ICBM strike, it is highly likely the president would consult with any country in the flight path, limiting overflight concerns as a constraint on action. In addition, the United States and Russia could develop an agreement of pre-notification to mitigate risk of Moscow misinterpreting a launch. Overflight could be avoided by choosing a suboptimal flight path and increasing the CEP, a trade-off that might be acceptable when national security is at risk.

Not only do ICBMs have needed physical attributes, but they also provide nuclear deterrence at a low cost. The 2014 Department of Defense budget was $618 billion; $12 billion was specifically budgeted for strategic (nuclear) forces.[63] The entire ICBM force cost less than $1 billion annually to operate and maintain—less than 0.02 percent of the total defense budget.[64] The cost is so low that removing them from the fleet will provide little real savings, contrary to assertions otherwise. ICBMs recently underwent modernization upgrades that will keep them operational at least through 2030.[65]

The Air Force is also working on the follow-on to the Minuteman III. Estimates suggest the future system will cost about $20–$70 billion.[66] This may seem like a large investment, but in actually it is less than 0.06 percent of the total defense budget over the length of the modernization activities. Additionally, since the system is land-based, without flight or at-sea costs, it should remain the least costly nuclear weapon system in the US nuclear arsenal. The figure below illustrates the extremely low costs of the ICBM relative to both the strategic forces as a whole, but in particular to the entire Department of Defense budget.

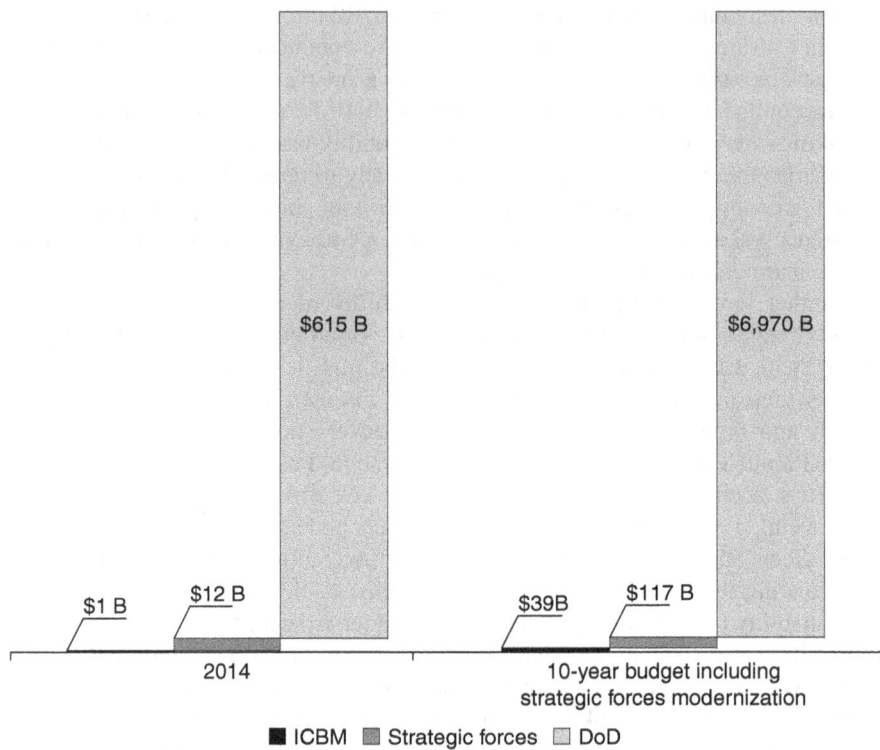

$1 B $12 B

$615 B

2014

$39B $117 B

$6,970 B

10-year budget including
strategic forces modernization

■ ICBM ▨ Strategic forces ☐ DoD

Figure 7.3 ICBM cost relative to strategic force and entire Department of Defense budget[67]

The ICBM's development to counter a Cold War threat does not make it irrelevant in today's security environment. America needs to be prepared to counter both traditional and emerging threats, while at the same time retaining the capability to defeat nations that today are not considered nuclear threats, but may become one in the future. There are no indications that adversaries intend to divest themselves of their nuclear weapons programs. Americans would be mistaken to believe that they or allies would be more secure without a reliable, flexible, and diversely survivable nuclear arsenal. Rather, the United States and her allies are best protected by a credible nuclear deterrent force that includes a highly competent ICBM component.

Recommendations

Modernize and recapitalize the ICBM fleet

The Minuteman III is currently undergoing life-extension programs to maintain viability through 2030. However, now is the time to begin working toward a next-generation ICBM. From a practical perspective, ensuring that a follow-on ICBM program becomes part of the research and development program of record is essential, given the difficult defense budgets ahead. Once employed, a new system is likely to prove of similar cost-effectiveness to the Minuteman III. While initial development costs will prove significant, as all weapon systems are, the long-term return on investment will make a next-generation ICBM well worth the expenditure. That being said, the physical infrastructure at the launch facilities are sound despite their age, and do not require many, if any, upgrades. The follow-on ICBM and control centers should be housed in the existing facilities, with an estimated savings of several billion dollars in modernization cost.[68] Minuteman III life-extension programs were designed to keep the system operational through 2020—trying to push it much past the 2030 time frame is asking for significantly decreased reliability, reducing the credibility of the nuclear force.

Do not reduce the number of ICBMs below 400

Under the current force structure, Russia will need to attack 400 ICBMs along with 45 launch control centers to destroy America's ICBM force, requiring about two-thirds of their warheads. Although the Department of Defense has seemingly decided to reduce the number of on-alert missiles to 400, moving below this number would challenge the credibility of the arsenal as it signals a lack of confidence in the ICBM force.[69] Further reductions in the ICBM force structure leaves America significantly more vulnerable to a debilitating attack and with only a limited second-strike capability.

To reduce the number of US-deployed ICBMs below 400 not only runs the risk of the United States being unable to respond to a Russian first strike (or coercion), it also reduces the confidence American allies have in the country's pledge to extend deterrence to them. In short, the smaller the US arsenal, the greater the chance of proliferation. Allies may view fewer nuclear weapons as little more than a reserve to protect the US homeland. This could drive America allies to develop their own nuclear arsenals. Adversaries with smaller arsenals may interpret further American reductions in ICBMs as a signal that there is an opportunity to become a nuclear peer, encouraging the development of additional weapons and systems. Going below 400 ICBMs not only makes the United States more vulnerable to attack, it also risks increasing proliferation by both friend and foe.

Do not reduce on-alert posture of ICBMs

Detractors believe the on-alert posture of the ICBM is destabilizing and could lead to mistaken launch. The 2010 *Nuclear Posture Review* (NPR) initiated studies to consider lowering the on-alert posture of ICBMs in the future (although the report did not recommend immediate changes to the alert posture).[70] To reduce the ICBM alert posture removes one of the key components of a reliable nuclear force, prompt response. The current posture keeps adversaries' targets at risk around the clock ensuring the United States will not be caught by surprise and unable to respond. It also gives the president as much time as possible to decide if a nuclear strike is required.[71] Without ICBMs on alert, the president may feel more rushed to make a decision knowing the time it takes to ready an ICBM force that is not on alert. Having ICBMs on continual alert ensures that both launch and maintenance crews maintain the proficiency needed to ensure flawless operations. Without a constant alert status, crews run a greater risk of error, possibly causing future issues with crew performance both while on missile alert or in the testing and evaluation environment; a problem that has already arisen in an environment where crews do not view their mission as critical to the nation's security. Finally, building in a delay leaves the United States and her allies vulnerable to an unexpected attack, possibly prompting allies to build their own force since they do not believe they can rely on the United States to act immediately.

Consider developing a small force of mobile ICBMs

The 2010 NPR recommended exploring basing options to maximize survivability of the ICBM force.[72] Although the United States should maintain and modernize the current fleet of 450 ICBMs, the Department of Defense should also explore the possibility of enhancing this force through a small fleet of mobile ICBMs. Mobile ICBMs would increase the number of targets, further complicating targeting for nuclear foes along with building in additional survivability together with the rapid response of on-alert ICBMs. Mobile ICBMs enhance survivability by being difficult to locate since they are not in a fixed location as well as having the ability to use natural barriers, such as mountainous terrain, as an added measure of protection from incoming missiles. Not only does a mobile ICBM force provide the president with the same on-alert posture as the current ICBM, but it also serves as a hedge against an adversary developing technology in the future which would allow easy detection of at-sea nuclear-armed submarines.

Conclusion

Weapons developed during the Cold War to combat a specific enemy, the Soviet Union, cannot be deemed obsolete or unnecessary just because the security environment is different. In reality, threats to the United States and her allies have changed and expanded, which makes the ICBM as relevant today as when the system was first conceived. Russia and other potential adversaries have shown no inclination to remove land-based nuclear weapons as a key component of their warfighting arsenal.

The United States, both for its own security and the security of its allies, needs ICBMs to ensure nuclear deterrence. However, in order for the weapon system to be an enduring and credible threat, the United States must modernize the force and keep current missiles on alert while exploring the possibility of increasing the weapon's survivability by adding a mobile ICBM. Without the ICBM, the United States will become a less diverse and flexible strategic nuclear force structure, especially in delivery systems for prompt attack, making it much more vulnerable to a nuclear attack and potentially paving the way toward greater nuclear proliferation by allies who no longer have faith the US nuclear umbrella will protect them. Nuclear weapons are not just for destruction, they are first and foremost to ensure deterrence.

Notes

1 Richard Smoke, *National Security and the Nuclear Dilemma* (Boston: McGraw Hill, 1993), 104–105.
2 Adam Lowther, *Challenging Nuclear Abolition* (Maxwell AFB, AL: Air Force Research Institute, August 2009), 6.
3 Smoke, *National Security and the Nuclear Dilemma*, 120.
4 Amy F. Woolf, *US Strategic Nuclear Forces: Background, Development and Issues*, Congressional Research Service Report (Washington, DC: US Congress, June 2013), 32.
5 United States Department of Defense, "Fact Sheet on U.S. Nuclear Force Structure under the New START Treaty," www.defense.gov/documents/Fact-Sheet-on-US-Nuclear-Force-Structure-under-the-New-START-Treaty.pdf.
6 United States Air Force, "LGM-30G Minuteman III," fact sheet, www.af.mil/AboutUs/FactSheets/Display/tabid/224/Article/104466/lgm-30g-minuteman-iii.aspx.
7 United States Air Force, "LGM-30G Minuteman III."
8 Global Zero, *Modernizing U.S. Strategy, Force Structure and Policy*, US Nuclear Policy Commission Report (Washington, DC: Global Zero, 2012), 8.
9 Harold A. Feiveson, *The Nuclear Turning Point: A Blueprint for Deep Cuts and De-alerting of Nuclear Weapons* (Washington, DC: Brookings Institute Press, 1999), 127.
10 Global Zero, *Modernizing U.S. Strategy, Force Structure and Policy*, 5.
11 Stephen J. Cimbala, "Nuclear Arms Reductions, Abolition and Nonproliferation: What's Ideal, What's Possible, What's Problematical," *Journal of Slavic Military Studies*, 22, 3 (July–September 2009), 329–351.
12 Benjamin Friedman, Christopher Preble, and Matt Fay, *The End of Overkill? Reassessing U.S. Nuclear Policy* (Washington, DC: CATO Institute, 2013), 10.
13 William M. Evans and Mark Manion, *Minding the Machines: Preventing Technical Disasters* (Upper Saddle River, NJ: Prentice Hall PTR, 2002), 45.
14 Global Zero, *Modernizing U.S. Strategy, Force Structure and Policy*, 7.
15 Scott D. Sagan and Kenneth Waltz, *The Spread of Nuclear Weapons* (New York: W.W. Norton, 2003), 36.
16 Dana J. Johnson, Christopher J. Bowie, and Robert P. Haffa, *Triad, Dyad, Monad? Shaping U.S. Nuclear Forces for the Future* (Arlington, VA: Air Force Association, 2009), 22.
17 Stephanie Spies, "Nuclear Triad: To Cut or Not To Cut?" Center for Strategic and International Studies, September 22, 2011, https://csis.org/blog/nuclear-triad-cut-or-not-cut.
18 Jeff Richardson, "Shifting from a Nuclear Triad to a Nuclear Dyad," *Bulletin of Atomic Scientists* (September/October 2009), 1–10.
19 The United States Senate has not ratified the CTBT, but the United States has observed a unilateral moratorium on nuclear explosive testing. See United States Department of State, "Comprehensive Nuclear Test-Ban Treaty (CTBT)," www.state.gov/t/avc/c42328.htm.

20 Christopher A. Preble, *From Triad to Dyad Nuclear Proliferation Update*, February 2010, CATO Institute, www.cato.org/sites/cato.org/files/pubs/pdf/npu_february2010. pdf.

21 There have been studies on using the ICBM in a conventional role, but it is generally accepted that due to concerns other nations would not be able to determine if a launch were conventional or nuclear until impact, the ICBM remains a nuclear-only weapon.

22 Defense Strategic Guidance, "Sustaining Global Leadership: Priorities for the 21st Century Defense," January 2012, 5.

23 Andrei Kokoshin, *Ensuring Strategic Stability in the Past and Present: Theoretical and Applied Questions*, Belfer Center for Science and International Affairs (Cambridge, MA: Harvard Kennedy School, 2011), 35.

24 James M. Kowalski, US Air Force, "Air Force Global Strike Command," briefing, September 4, 2013. Systems listed may be conventional with a potential nuclear capability and are not all inclusive. Above systems represent significant new developments. Musudan and TD-2 are of unknown operational status.

25 Franklin C. Miller, *The Need for a Strong US Nuclear Deterrent in the 21st Century* (Washington DC: Submarine Industrial Base Council, 2013), 5.

26 Robert Gates, Secretary, Department of Defense, "Gates: Nuclear Weapons and Deterrence in the 21st Century," address, Carnegie Endowment for International Peace, Washington, DC, October 28, 2008.

27 Mark B. Schneider, National Institute for Public Policy, "The Nuclear Forces and Doctrine of the Russian Federation and the People's Republic of China." testimony before the Armed Services Subcommittee on Strategic Forces, US House of Representatives, October 14, 2011.

28 Yury E. Fedorov, *New Wine in Old Bottles? The New Salience of Nuclear Weapons*, Security Studies Center and the Atomic Energy Commission, Proliferation Papers, Fall 2007, 18.

29 Congressional Commission on the Strategic Posture of the United States, *America's Strategic Posture: The Final Report of the Congressional Commission on the Strategic Posture of the United States* (Washington, DC: US Institute of Peace Press, 2009), 12; Harry de Quetteville and Andrew Pierce, "Russia Threatens Nuclear Attack on Poland over US Missile Shield Deal," *The Telegraph*, August 15, 2006, www.telegraph.co.uk/news/worldnews/europe/russia/2566005/Russia-threatens-nuclear-attack-on-Poland-over-US-missile-shield-deal.html; and John Newhouse, "Diplomacy, Inc.: The Influence of Lobbies on U.S. Foreign Policy," *Foreign Affairs*, 88, 3 (May/June 2009), 73–92.

30 Schneider, testimony; and Johnson, Bowie, and Haffa, *Triad, Dyad, Monad?* 9.

31 Robert M. Gates, *From the Shadows: The Ultimate Insider's Story of Five Presidents and How They Won the Cold War* (New York: Simon and Schuster, 1996), 570.

32 However, there is some question whether Russia's economy and underperforming military-industrial complex can fulfill the ambitious modernization program set forth by Putin and the Defense Ministry.

33 Arms Control Association, "Nuclear Weapons: Who Has What at a Glance," www. armscontrol.org/factsheets/Nuclearweaponswhohaswhat; and Johnson, Bowie, and Haffa, *Triad, Dyad, Monad?* 9.

34 Elbridge A. Colby and Abraham M. Denmark, "Nuclear Weapons and U.S.–China Relations: A Way Forward," Center for Strategic Studies, PONI working group on US–China nuclear dynamics, March 2013, 6, 12.

35 Schneider, testimony.

36 Rebeccah Heinrichs, *China's Strategic Capabilities and Intent*, Heritage Foundation *Issue Brief* No. 4111, December 18, 2013, www.heritage.org/research/reports/2013/12/ china-s-strategic-forces-military-capabilities-and-intent; and *Nuclear Posture Review Report* (Washington, DC: Department of Defense, 2010), x.

37 Miller, *The Need for a Strong US Nuclear Deterrent in the 21st Century*, 2.
38 Zhao Yanrong, "Kim Seeks ROK Ties, Warns US of Nuclear Disaster," *China Daily USA*, January 2, 2014, http://usa.chinadaily.com.cn/epaper/2014-01/02/content_1721 1044.htm.
39 Arms Control Association, "Nuclear Weapons: Who Has What at a Glance."
40 Arms Control Association, "Nuclear Weapons: Who Has What at a Glance."
41 Senate ICBM Coalition, *The Long Pole of the Nuclear Umbrella* (Washington, DC: US Senate ICBM Coalition, November 2009), 10; and Congressional Commission on the Strategic Posture of the United States, *America's Strategic Posture*, 69.
42 Nuclear Accidents/Incidents, www.9websites.com/airforce/nucacc.htm.
43 Miller, *The Need for a Strong US Nuclear Deterrent in the 21st Century*, 8.
44 Senate ICBM Coalition, *The Long Pole of the Nuclear Umbrella*, 5.
45 I. C. Oelrich, *Sizing Post-Cold War Nuclear Forces* (Alexandria, VA: Institute For Defense Analyses, October 2001), 45.
46 William J. Perry and Brent Scowcroft, *US Nuclear Weapons Policy*, Independent Task Force Report No. 62 (New York: Council on Foreign Relations, 2009), 35.
47 Congressional Commission on the Strategic Posture of the United States, *America's Strategic Posture*, 22.
48 Charles V. Peña, "Strategic Nuclear Forces and Missile Defense," in *CATO Handbook for Policymakers* (Washington, DC: CATO Institute, September 2005), 525.
49 Colin S. Gray, "The Strategic Triad: End of the Road?" *Foreign Affairs*, 56, 4 (July 1978): 771–789.
50 Lowther, *Challenging Nuclear Abolition*, 14.
51 Glenn Buchan, David Matonick, Calvin Shipbaugh, and Richard Mesic, *Future of U.S. Nuclear Forces* (Washington, DC: RAND Corp, 2005).
52 Senate ICBM Coalition, *The Long Pole of the Nuclear Umbrella*, 11. Gen. Larry D. Welch, US Air Force, retired, "Deterrence and the Triad," address, Kings Bay, GA, November 7, 2013.
53 *Nuclear Posture Review Report*, ix.
54 Woolf, *US Strategic Nuclear Forces*, 11.
55 US Department of State, "New START Treaty," www.state.gov/t/avc/newstart/index. htm.
56 Preble, *From Triad to Dyad*.
57 Kowalski, "Air Force Global Strike Command."
58 Air Force Global Strike Command Instruction 99–102, *Intercontinental Ballistic Missile Operational Test and Evaluation*, March 2, 2011, 5; and US Strategic Command briefing, www.dtic.mil/ndia/2011SET/TYNER.pdf.
59 Darci Bloyer and Zachariah M. Becker, *Building an Extended U.S. Nuclear Deterrent for the 21st Century*, CSIS Project on Nuclear Issues, Center for Strategic and International Studies, 13–15http://csis.org/images/stories/poni/110921_Bloyer_and_Becker. pdf.
60 Woolf, *US Strategic Nuclear Forces*, 13.
61 Circular error probable (CEP) is a measure of a ballistic missile's accuracy used to determine probable damage of a target. CEP is the radius of a circle in which a missile has a 50 percent chance of hitting.
62 James Dunnigan, *Digital Soldiers: The Evaluation of High Tech Weaponry and Tomorrow's Brave New Battlefield* (New York: St. Martin's Press, 1996), 156.
63 The White House, "Department of Defense Budget," www.whitehouse.gov/sites/ default/files/omb/budget/fy2014/assets/defense.pdf.
64 Peter Huessy, "In Defense of the Nuclear Triad," Defense One, October 18, 2013, www.defenseone.com/ideas/2013/10/defense-nuclear-triad/72242/?oref=d-interstitial-continue.
65 John Shaud, *In Service to the Nation: Air Force Research Institute Strategic Concept for 2018–2023* (Maxwell AFB, AL: Air University Press, 2009), 35.

66 Jon B. Wolfsthal, Jeffrey Lewis, and Marc Quint, "The Trillion Dollar Nuclear Triad: US Strategic Nuclear Modernization Over the Next Thirty Years," James Martin Center for Nonproliferation Studies, January 2014, 22.
67 Wolfsthal et al., "The Trillion Dollar Nuclear Triad: US Strategic Nuclear Modernization Over the Next Thirty Years"; and The White House, "Department of Defense Budget."
68 Wolfsthal et al., "The Trillion Dollar Nuclear Triad," 22.
69 US Department of Defense, "Fact Sheet on U.S. Nuclear Force Structure."
70 *Nuclear Posture Review Report*, 26.
71 US Air Force, "Flight Plan for the Nuclear Enterprise," June 26, 2013, 6.
72 *Nuclear Posture Review Report*, x.

Bibliography

Arms Control Association, "Nuclear Weapons: Who Has What at a Glance," www.armscontrol.org/factsheets/Nuclearweaponswhohaswhat.

Bloyer, Darci, and Becker, Zachariah M., *Building an Extended U.S. Nuclear Deterrent for the 21st Century*, CSIS Project on Nuclear Issues, Center for Strategic and International Studies, 13–15http://csis.org/images/stories/poni/110921_Bloyer_and_Becker.pdf.

Cimbala, Stephen J., "Nuclear Arms Reductions, Abolition and Nonproliferation: What's Ideal, What's Possible, What's Problematical," *Journal of Slavic Military Studies*, 22, 3 (July–September 2009).

Colby, Elbridge A., and Denmark, Abraham M., "Nuclear Weapons and U.S.–China Relations: A Way Forward," Center for Strategic Studies, PONI working group on US–China nuclear dynamics, March 2013.

Congressional Commission on the Strategic Posture of the United States, *America's Strategic Posture: The Final Report of the Congressional Commission on the Strategic Posture of the United States* (Washington, DC: US Institute of Peace Press, 2009).

Defense Strategic Guidance, "Sustaining Global Leadership: Priorities for the 21st Century Defense," January 2012.

de Quetteville, Harry, and Pierce, Andrew, "Russia Threatens Nuclear Attack on Poland over US Missile Shield Deal," *The Telegraph*, August 15, 2006, www.telegraph.co.uk/news/worldnews/europe/russia/2566005/Russia-threatens-nuclear-attack-on-Poland-over-US-missile-shield-deal.html.

Dunnigan, James, *Digital Soldiers: The Evaluation of High Tech Weaponry and Tomorrow's Brave New Battlefield* (New York: St. Martin's Press, 1996).

Evans, William M., and Manion, Mark, *Minding the Machines: Preventing Technical Disasters* (Upper Saddle River, NJ: Prentice Hall PTR, 2002).

Fedorov, Yury E., *New Wine in Old Bottles? The New Salience of Nuclear Weapons*, Security Studies Center and the Atomic Energy Commission, Proliferation Papers, Fall 2007.

Feiveson, Harold A., *The Nuclear Turning Point: A Blueprint for Deep Cuts and De-alerting of Nuclear Weapons* (Washington, DC: Brookings Institute Press, 1999).

Friedman, Benjamin, Preble, Christopher, and Fay, Matt, *The End of Overkill? Reassessing U.S. Nuclear Policy* (Washington, DC: CATO Institute, 2013).

Gates, Robert, Secretary, Department of Defense, "Gates: Nuclear Weapons and Deterrence in the 21st Century," address, Carnegie Endowment for International Peace, Washington, DC, October 28, 2008.

Gates, Robert M., *From the Shadows: The Ultimate Insider's Story of Five Presidents and How They Won the Cold War* (New York: Simon and Schuster, 1996).

Global Zero, *Modernizing U.S. Strategy, Force Structure and Policy*, US Nuclear Policy Commission Report (Washington, DC: Global Zero, 2012).

Gray, Colin S., "The Strategic Triad: End of the Road?" *Foreign Affairs*, 56, 4 (July 1978).

Heinrichs, Rebeccah, *China's Strategic Capabilities and Intent*, Heritage Foundation *Issue Brief* No. 4111, December 18, 2013, www.heritage.org/research/reports/2013/12/china-s-strategic-forces-military-capabilities-and-intent.

Huessy, Peter, "In Defense of the Nuclear Triad," Defense One, October 18, 2013, www.defenseone.com/ideas/2013/10/defense-nuclear-triad/72242/?oref=d-interstitial-continue.

Johnson, Dana J., Bowie, Christopher J., and Haffa, Robert P., *Triad, Dyad, Monad? Shaping U.S. Nuclear Forces for the Future* (Arlington, VA: Air Force Association, 2009).

Kokoshin, Andrei, *Ensuring Strategic Stability in the Past and Present: Theoretical and Applied Questions*, Belfer Center for Science and International Affairs (Cambridge, MA: Harvard Kennedy School, 2011).

Kowalski, James M., US Air Force, "Air Force Global Strike Command," briefing, September 4, 2013.

Lowther, Adam, *Challenging Nuclear Abolition* (Maxwell AFB, AL: Air Force Research Institute, August 2009).

Miller, Franklin C., *The Need for a Strong US Nuclear Deterrent in the 21st Century* (Washington DC: Submarine Industrial Base Council, 2013).

Newhouse, John, "Diplomacy, Inc.: The Influence of Lobbies on U.S. Foreign Policy," *Foreign Affairs*, 88, 3 (May/June 2009).

Nuclear Accidents/Incidents, www.9websites.com/airforce/nucacc.htm.

Nuclear Posture Review Report (Washington, DC: Department of Defense, 2010).

Oelrich, I. C., *Sizing Post-Cold War Nuclear Forces* (Alexandria, VA: Institute For Defense Analyses, October 2001).

Peña, Charles V., "Strategic Nuclear Forces and Missile Defense," in *CATO Handbook for Policymakers* (Washington, DC: CATO Institute, September 2005).

Perry, William J., and Scowcroft, Brent, *US Nuclear Weapons Policy*, Independent Task Force Report No. 62 (New York: Council on Foreign Relations, 2009).

Preble, Christopher A., *From Triad to Dyad Nuclear Proliferation Update*, February 2010, CATO Institute, www.cato.org/sites/cato.org/files/pubs/pdf/npu_february2010.pdf.

RAND: Project Air Force, *Future of U.S. Strategic Forces*, October 14, 2011.

Richardson, Jeff, "Shifting from a Nuclear Triad to a Nuclear Dyad," *Bulletin of Atomic Scientists* (September/October 2009).

Sagan, Scott D., and Waltz, Kenneth, *The Spread of Nuclear Weapons* (New York: W.W. Norton, 2003).

Shaud, John, *In Service to the Nation: Air Force Research Institute Strategic Concept for 2018–2023* (Maxwell AFB, AL: Air University Press, 2009).

Schneider, Mark B., "The Nuclear Forces and Doctrine of the Russian Federation and the People's Republic of China," National Institute for Public Policy, testimony before the Armed Services Subcommittee on Strategic Forces, US House of Representatives, October 14, 2011.

Senate ICBM Coalition, *The Long Pole of the Nuclear Umbrella* (Washington, DC: US Senate ICBM Coalition, November 2009).

Smoke, Richard, *National Security and the Nuclear Dilemma* (Boston: McGraw Hill, 1993).

Spies, Stephanie, "Nuclear Triad: To Cut or Not To Cut?," Center for Strategic and International Studies, September 22, 2011, https://csis.org/blog/nuclear-triad-cut-or-not-cut.

US Air Force, "Flight Plan for the Nuclear Enterprise," June 26, 2013.

US Air Force, "LGM-30G Minuteman III" fact sheet, www.af.mil/AboutUs/FactSheets/Display/tabid/224/Article/104466/lgm-30g-minuteman-iii.aspx.

United States Air Force Global Strike Command Instruction 99–102, *Intercontinental Ballistic Missile Operational Test and Evaluation*, March 2, 2011, 5; and US Strategic Command briefing, www.dtic.mil/ndia/2011SET/TYNER.pdf.

United States Department of Defense, "Fact Sheet on U.S. Nuclear Force Structure under the New START Treaty," www.defense.gov/documents/Fact-Sheet-on-US-Nuclear-Force-Structure-under-the-New-START-Treaty.pdf.

United States Department of State, "Comprehensive Nuclear Test-Ban Treaty (CTBT)," www.state.gov/t/avc/c42328.htm.

United States Department of State, "New START Treaty," www.state.gov/t/avc/newstart/index.htm.

Welch, Gen. Larry D., US Air Force, retired, "Deterrence and the Triad," address, Kings Bay, GA, November 7, 2013.

White House, "Department of Defense Budget," www.whitehouse.gov/sites/default/files/omb/budget/fy2014/assets/defense.pdf.

Wolfsthal, Jon B., Lewis, Jeffrey, and Quint, Marc, "The Trillion Dollar Nuclear Triad: US Strategic Nuclear Modernization Over the Next Thirty Years," James Martin Center for Nonproliferation Studies, January 2014.

Woolf, Amy F., *US Strategic Nuclear Forces: Background, Development and Issues*, Congressional Research Service Report (Washington, DC: US Congress, June 2013).

Yanrong, Zhao, "Kim Seeks ROK Ties, Warns US of Nuclear Disaster," *China Daily USA*, January 2, 2014, http://usa.chinadaily.com.cn/epaper/2014-01/02/content_1721 1044.htm.

8 Strategic ballistic missile submarines

A necessary but uncertain future

Donald M. Neff

Background

Since 1960, when the USS *George Washington* began its first Polaris armed nuclear deterrence patrol, strategic ballistic missile submarines (SSBNs) have served as a pivotal element in the United States's nuclear triad. For the 44 years of the Cold War, the SSBN brought unique capabilities to the nuclear deterrence mission and helped ensure American forces survivability, mobility, and credibility. After the fall of the Soviet Union precipitated drastic reductions in America's nuclear arsenal, the viability of the nuclear triad of bombers, intercontinental ballistic missiles (ICBMs), and submarine-launched ballistic missiles (SLBMs) was called into question by academics and policymakers within and without the military who advocated eliminating one or more leg of the triad and changing the nation's nuclear posture.

In his 2009 Prague speech, President Barack Obama, while committed to a world without nuclear weapons, said, "As long as these weapons exist, the United States will maintain a safe, secure and effective arsenal to deter any adversary, and guarantee that defense to our allies."[1] The following year the administration's *Nuclear Posture Review* (NPR) concluded that the US nuclear triad would be maintained under the New Strategic Arms Reduction Treaty (New START) agreement, which limited America's arsenal to 1,550 operationally deployed strategic nuclear weapons.[2] In his 2013 Berlin speech, the president committed to negotiating a further reduction of Russian and American arsenals of up to one-third.[3] While criticized by Russian leaders, Obama's comments underpin momentum to reduce the role played by America's nuclear force.[4] Continued calls for further force reductions amid persistent Global Zero lobbying efforts and significant fiscal pressures will challenge triad proponents to offer a clear defense of the arsenal as legacy weapon systems begin to reach the end of their service life expectancies.

Despite a volatile and uncertain fiscal and strategic environment, the current Ohio-class SSBN and its planned replacement (SSBNx) will remain crucial elements of the United States's strategic nuclear deterrent. The unique capacities that the SSBN brings to the nuclear triad may not, however, be enough to counter the increasing pressure to reduce the nuclear arsenal and divest expensive weapon systems. The following pages will review the current and future SSBN force structure, plans, and time lines; assess arguments against and for the SSBN; and evaluate mission-expanding options that have potential to transform the SSBN from the silent Cold War weapon of its past to a flexible, projectable, and employable strike option to deter and counter current and future security threats.

Force structure

As of January 2015, the US had about 1,650 strategic warheads deployed on a mix of about 1,000 SLBMs, bombers, and ICBMs.[5] Fourteen Ohio-class SSBNs remain in nuclear deterrence service and will be increasingly relied on as the nation's key strategic nuclear force. As the most survivable leg of the nuclear triad, the SSBN and Trident II D-5 SLBMs will comprise the majority of America's operationally deployed strategic nuclear weapons.[6]

With 14 operational SSBNs in the fleet, two submarines are rotated through mid-life refuel and maintenance at any given time and are therefore not counted against New START numbers. Under the new treaty, four of the 24 trident missile launchers on each submarine will be eliminated reducing the total number of deployed Trident launchers to 240 and total nuclear warheads to no more than 1,090. The total number of warheads will depend on the number of multiple independently targetable reentry vehicles (MIRV) loaded on each missile.[7]

In October 2004, the last of the Trident I (C-4) missiles were removed from the SSBN fleet and replaced with the larger and more accurate Trident II (D-5) SLBM. The increased range of the D-5 allows SSBNs to expand their potential patrol areas, further complicating opponents' defenses. Actual combat range of the three-stage, 130,000-pound D-5 is classified and depends largely on payload weight. Effective missile range is estimated, however, to be in excess of 4,000 nautical miles.[8] While the D-5 is capable of carrying up to eight independently targetable reentry vehicles, under New START counting rules the US can download, customize, and tailor MIRV packages for each missile depending on target characteristics and ranging requirements. The US will only be required to disclose the total number of warheads deployed, not how many weapons are loaded on each missile.[9]

To improve targeting options against potential opponents in Asia, nine Ohio-class SSBNs are based at Naval Base Kitsap Bangor, Washington, with the remaining five stationed at Kings Bay, Georgia.[10] With each base having one boat continuously undergoing refuel and maintenance, 12 submarines (eight in the Pacific Ocean and four in the Atlantic Ocean) are operational at all times with approximately half on nuclear deterrence patrol.

Ohio-class submarines became operational between 1981 and 1997 with an extended 42-year service life comprised of two 20-year operational periods divided by a two-year refuel/overhaul process.[11] In 2027, the first Ohio-class will reach the end of its service life with the remaining submarines timing out at approximately one per year with full retirement occurring in 2040 (see Figure 8.1). To match the life expectancy of Ohio-class submarines and serve as the initial SLBM for the follow-on SSBNx, the Trident II D-5 and mated W76/88 reentry vehicles will be updated through programmed life-extension efforts.[12] By the end of 2029 the first two Ohio-class SSBNs will have reached the end of their service lives and the follow-on replacement SSBNx will need to become operational to replace the retiring boats without affecting continuous nuclear deterrence patrols.[13]

The US Navy (USN) plans on procuring 12 SSBNx to replace the 14 Ohio-class submarines with the first SSBNx becoming operational in 2030 and the last in 2042 (see Figure 8.1).[14] The SSBNx is expected to be more sustainable and less prone to maintenance issues than Ohio-class vessels. Therefore, according to the Navy, only ten operational SSBNx boats will be needed to ensure nuclear deterrence.[15]

The new submarine will have the same 40-year life expectancy but not require the lengthy two-year mid-life refueling overhaul plan as the Ohio-class boats.[16] Combining the current procurement and retirement plans, the Navy acknowledges the risk of reducing the total number of operational ballistic missile submarines to just ten between 2032 and 2040 but mitigates the problem by not having any Ohio-class boats undergoing mid-life overhaul during those years. The life-extended Trident II D-5 will be the initial SLBM deployed on the SSBNx and each submarine will carry 16 missiles verses the 20 carried on the latest modified Ohio-class.[17] After initial engineering costs are incurred, the Navy plans to limit the expense of each SSBNx to $4.9 billion per submarine, with construction of the first submarine slated for 2021.[18]

Arguments against SSBNs

With the continued reductions of strategic nuclear inventories in line with the Russian–American New START treaty and reduction goals made by President

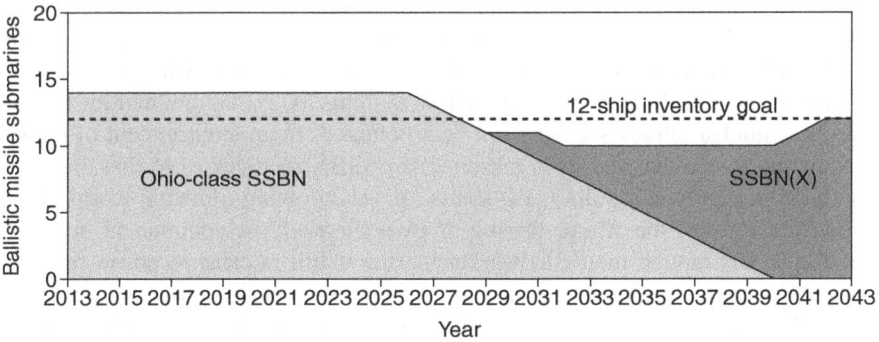

Figure 8.1 US Navy's 2014 30-year SSBN building plan

Source: CBO Analysis of Navy's FY2014 Shipbuilding Plan.

Obama in his Berlin speech, the strategic missile submarine will continue to be relied on as a safe and secure projector of America's nuclear forces. However, detractors make several arguments against the development and employment of SSBNs.

Disarmament proponents contend that the unique stealthy nuclear-strike characteristics presented by the SSBN, in addition to advancements in worldwide anti-ballistic missile (ABM) technology and conventional prompt global-strike weapons is creating a first-strike capability targeted against any adversary.[19] So-called first-strike capability is where a disarming first attack against an enemy's forces can be achieved to the degree that the enemy is unable to strike back with any effectiveness.[20]

It is argued that the United States's emerging conventional and nuclear sword and shield will be comprised of the traditional nuclear triad as well as a globally deployed network of anti-ballistic missile interceptors capable of neutralizing an opponent's retaliatory nuclear response. This strategic buildup coupled to the Pentagon's "Pivot to Asia" is argued to be a destabilizing force in the Pacific region, which may encourage China and Russia to form stronger diplomatic and military ties to counterbalance the United States.[21]

The ability of Ohio-class ballistic missile submarines, carrying the Trident D-5 SLBM, patrolling off the coasts of China, Russia, and North Korea striking targets with minimal detection and flight time threaten to destabilize the nuclear balance of power. As argued by these opponents, the absolute numbers of weapons needed for effective nuclear balance and deterrence is less relevant than the ensured capability to retaliate. As long as balancing nations have adequate second-strike capability, strategic deterrence is likely to be achieved.[22] If improvements to the current and future SSBN force, coupled with ABM and counterforce/denial capabilities emerge, the second-strike capability of America's adversaries will be neutralized and strategic deterrence will be threatened, resulting in a destabilized strategic environment and potential conflict and arms proliferation.

An enduring argument, originally posed by RAND Corporation analysts, against the fielding and deployment of SSBNs suggests there is a threat of well-intentioned but unauthorized use of nuclear weapons.[23] Because of the remote and concealed nature of SSBN employment and reliance on slow and less robust command, control, and communication systems (C3), an unintentional catastrophic nuclear attack scenario can be articulated. In an environment of rapidly escalating world tensions, it is reasoned, an SSBN receives word that the US is under attack at which point C3 systems fail. Unknowing of actual world events and the nature of the attack, fearing its own survival, the commander and crew of the SSBN launch their SLBMs, triggering a full nuclear response from the targeted state and in-kind retaliation by the United States. With this logic, one of the SSBN's greatest strengths (the autonomous and decoupled nature of SSBN nuclear deterrence operations) exposes and increases unwarranted risk of unintended and unauthorized nuclear weapons release and is thus inherently dangerous and destabilizing.

The pivotal role of SSBNs in America's nuclear triad is to maintain a credible and survivable second-strike capability ensuring balanced strategic deterrence, regardless of enemy nuclear forces or capabilities.[24] However, the actual utility and credibility of these forces can be called into question. If deterrence fails, the nature of a second strike against an opponent, in response to their first use, may be a punishing counter-value attack against that nation's cities and populations.[25] This argument reasons that counterforce targeting against enemy defenses and strategic weapons in a second-strike attack would be ineffective, since enemy missiles and bombers would have already launched. Attacking empty missile silos and vacant airfields with retaliatory SLBMs would be fruitless. Following this logic, the only credible threat behind the SSBNs' second strike is that the SLBMs can target enemy cities and population centers, holding what the enemy values at risk of destruction. Two critical questions emerge that challenge the deterrent validity of the SSBN's second-strike capability. First, does the enemy *value* what is targeted enough to dissuade them from launching a first strike? Second, would an American president, faced with eminent destruction of the United States, authorize a massive counter-value, counter-population attack that potentially kills millions of people—knowing that the decision to attack was not that of the citizenry but the government? If either answer is no, then the utility and credibility of the SSBN as a second-strike platform is questionable. Since the United States has never adopted a no-first-use nuclear policy, the limited and less credible counterforce second-strike utility of the SSBN bolsters the opponents' argument that nuclear-armed submarines are destabilizing first-strike weapons.

The second-strike capability and resulting deterrence provided by the SSBN is only applicable to adversaries with a credible first-strike capability, namely Russia and China. Lesser powers such as North Korea or a nuclear-armed Iran that decide to launch a nuclear attack against the United States or its allies would not have the ability to cripple North American-based retaliatory forces and would face the prospects of total destruction from continental assets. Whether the American counter-strike would remain conventional or escalate to nuclear is less important to the role played by the SSBN as SLBMs would not necessarily be required for retaliation. Further, any nuclear exchange with Russia or China would likely be an escalation of a conventional conflict where all US nuclear forces are at heightened levels of readiness, prepared to launch prior to an enemy's decapitating first strike.

Although inaccurate, it is possible to conclude from this discussion that the only utility SSBNs provide is against an unlikely attempt at a surprise decapitating counterforce first strike launched by Russia or China. The costs of developing and fielding the Ohio-class replacement will certainly compete for limited resources with the Navy's conventional forces. Should fiscal circumstances deteriorate further, a reduction in the projected SSBNx program could jeopardize the credibility of the seaborne leg of the nuclear triad. However, should the Navy develop the SSBNx at the expense of a conventional shipbuilding plan, the United States could be pressed into relying more heavily on less flexible nuclear deterrence to maintain security and stability in the international system.[26]

The Navy's 2014 shipbuilding plan and requested budget averages $21.2 billion per year for the next 30 years.[27] This planned outlay, derived by the Congressional Budget Office (CBO), is 13 percent higher than the Navy's $18.7 billion estimate and 34 percent higher than average amounts the Navy has received in recent decades (see Figure 8.2).[28] Between 2014 and 2043 the Navy plans on procuring 266 ships for a CBO calculated total cost of $636 billion in 2013 dollars (see Figure 8.3).

Generating an accurate cost estimate of fielding the 12 planned SSBNx is highly problematic and uncertain. In 2008 the Navy estimated the per-submarine cost at $3.8 billion but revised that number up in 2011 to $7.9 billion, a 108 percent increase in three years. The CBO estimates the total SSBNx cost to be between $97 billion and $102 billion or $7.2 billion per submarine plus between $10 billion and $15 billion in research and development costs for the program.[29]

At $100 billion, the SSBNx accounts for 15.7 percent of the Navy's total 30-year shipbuilding budget but only represents 4.5 percent of the desired ships; 12 of 266 hulls planned. As one of the Navy's highest procurement priorities, dedicating disproportionate funds to the SSBNx will be at the expense of lesser ranked combatant ships, calling into question US conventional naval requirements.[30] Compounding cost inaccuracies, the CBO evaluates that between 2014 and 2021 the Department of Defense underestimates its Future Years Defense Program (FYDP) budget by $209 billion.[31] The CBO also estimates that the Navy acquisition budget requirement will jump 10 percent in 2018 and an additional 16 percent in 2019 (see Figure 8.4).[32] In an environment of shrinking federal budgets and automatic spending cuts forced by the Budget Control Act

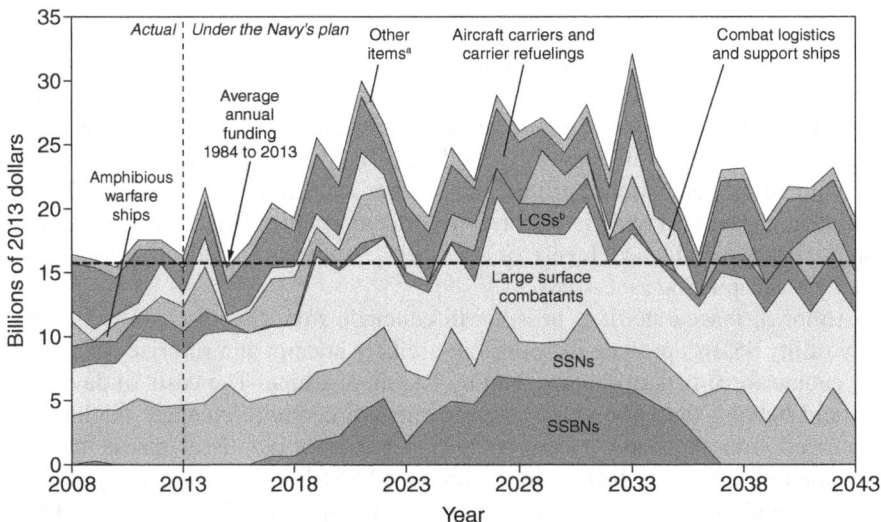

Figure 8.2 New ship construction annual costs—CBO estimates

Source: CBO Analysis of Navy's FY2014 Shipbuilding Plan.

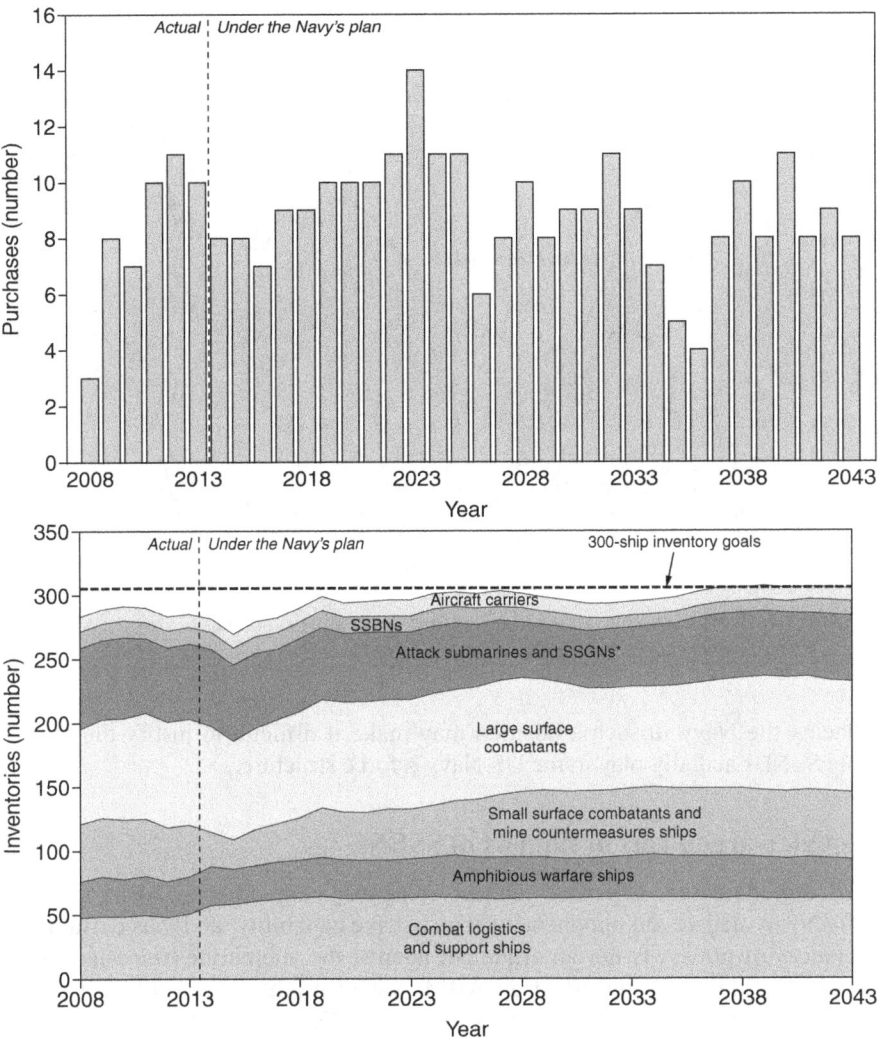

Figure 8.3 US Navy's 2014 30-year annual ship purchase and inventory plan

Source: CBO Analysis of Navy's FY2014 Shipbuilding Plan.

Note
* The Navy does not plan to build more SSGNs. The four Ohio-class SSGNs are expected to remain in service through the mid-2020s.

and resulting sequestration, it is unrealistic for the Navy to assume its 30-year shipbuilding plan will be fully funded. Allocating disproportionate funding to SSBNx could force a reduced conventional naval capability and decrease projectable US power (quantity has a quality all its own) and could—as mentioned—drive an increased reliance on less credible, low-utility nuclear forces.

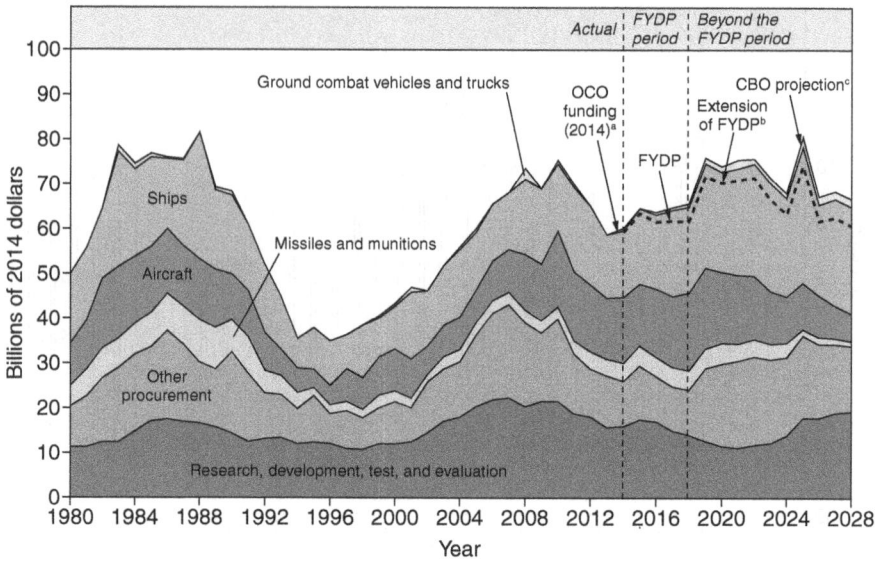

Figure 8.4 Costs of the Navy and Marine Corps's acquisition plans
Source: CBO Long-Term Implications of the 2014 FYDP.

Placing the Navy in such a position may make it difficult to justify the limited role SSNBs actually play in the US Navy's force structure.

Counter-arguments in support of SSBNs

The argument that offensive nuclear weapons, coupled with ABM systems, effectively negates an opponent's second-strike capability, and thus destabilizes balances of power, is not an argument against the submarine-based leg of the triad. Any destabilizing effect an ABM system might pose would not necessarily rely on submarine-based missiles as its offensive arm. Operating under a protective shield, land- or air-based, weapons could execute a first strike while retaliatory missiles are intercepted by an ABM system. Following this logic, any comprehensive ABM system would ensure first-strike survival of more cost-effective land-based ICBMs, thus eliminating the need for SLBMs entirely.

However, this argument is faulty because no ABM system is likely to be 100 percent effective. Despite shortened flight times, sea-launched SLBMs are unlikely to be capable of a crippling counter-force first strike. Both China and Russia have developed and deployed highly mobile ICBM systems which, while capable of being targeted with maneuverable high-yield weapons, are difficult to locate and continuously track.[33] For example, as seen in the Gulf War (1991), despite having complete air supremacy and an enduring presence, the United

States was unable to locate or destroy Iraqi Scud missiles targeted and launched against Israel.[34] Despite dramatic improvements in intelligence, surveillance, and reconnaissance technology since 1991, given a non-permissive environment, extensive geography, enemy countermeasures, and the extreme consequences of failed interdiction, it is highly unrealistic that national decision makers would contemplate a first strike in expectation of destroying most mobile- and silo-based launchers.

The most conclusive argument against viewing SSBNs as first-strike weapons lies in the number of deployed missiles and warheads. Given the reduced estimate of up to 1,090 total SLBM warheads, re-posturing of the SSBN force to nine boats in the Pacific and five in the Atlantic, and the common practice to have half the vessels on deterrent patrol, approximately 467 warheads would be available for an American SSBN surprise nuclear first strike.[35] This number of SLBM warheads is inadequate to carry out a wholly crippling first strike given that New START limits restrict deployed strategic inventories to 1,550 and Russian and Chinese SSBN deterrence patrols could not be effectively targeted. Increasing the number of American submarines on patrol to boost the number of reentry vehicles and/or generating North American-based nuclear forces would signal an increase of the United States's nuclear posture to adversaries, thus hampering the element of surprise and triggering a countering increase of enemy nuclear forces posture.

While an enduring and popular argument for authors and movie makers, the idea that a rogue SSBN commander would launch an unauthorized nuclear strike is highly problematic and unlikely. Robust and redundant communication systems coupled with clear command-and-control protocols all but eliminates potential unauthorized release. Physical and procedural permissive action links dramatically reduce risk and are proven safeguards that have secured the United States's nuclear inventory for many decades.[36] A survivable, reliable, and secure top-down Nuclear Command and Control System (NCCS) ensures the sole authority for nuclear employment and termination is retained by the president.[37] Absent definitive nuclear release authority issued by the president, warheads, weapons, and fire control systems are locked preventing unauthorized use.

While Russia and China pose a first-strike existential threat to the United States, lesser adversarial nations, while theoretically able to execute a nuclear strike against the United States, do not have the capacity to threaten all continental-based retaliatory forces. Thus the submarine-based deterrent force is not fully directed toward these lesser states or third-party non-state actors, although it can be on short notice. The two questions posed earlier are initial points to evaluate the effectiveness of SSBN second-strike credibility. First, does an adversary sufficiently value the targets of the SLBM enough to deter a first-strike attempt? And second, would an American president authorize a counter-value nuclear response knowing the fate of the United States is largely set? It is argued that all of Russia's or China's strategic missiles would have been launched during the initial attack, therefore SLBMs would target population centers with limited strategic significance.

It is incorrect to argue that significant strategic targets would not remain for an American retaliatory strike. Airfields, naval bases, aircraft carriers, munitions depots, petroleum reserves, fielded military forces, and, most critically, leadership and command-and-control nodes, would all still exist for American SLBMs to strike. These viable strategic centers of gravity undermine the foundation of the second question, namely that the president would be authorizing strikes against Russian and/or Chinese population centers rather than counterforce targets. Related to this point, the first question relies on the assumption that SLBMs would only target the less valued—by the Russian and Chinese governments—enemy population. To take this view requires a somewhat risky interpretation of what Russian and Chinese leaders value. If the highly probable assumption is that the regimes are most interested in self-preservation then an assured wholly devastating nuclear attack on military and governmental targets would provide sufficient deterrence preventing an initial attack. At the heart of nuclear deterrence is the mindset of the targeted opponent. The calculated ambiguity of what an American president might target with the secure SSBN second-strike force adds to enemy uncertainty and complicates a decision. It is this uncertainty, increased risk, and ambiguity that underpins the foundation of a second strike's assured destruction and nuclear deterrent balance.

The most convincing argument furthered by opponents of the SSBN is the high cost of the program and uncertainty related to defense acquisitions.[38] The Government Accountability Office's estimated 34 percent funding increase from historical levels is highly improbable given the declining fiscal environment for defense spending. To fully fund the development and fielding of the SSBNx under historically consistent budgeting, the Navy will be forced to curtail or eliminate portions of the total shipbuilding plan. Assuming forecasted procurement costs are accurate and the Navy receives historic funding, either fewer ships can be built or construction plans must be pushed further into the future to equalize the fiscal imbalance. Another option that should be considered is to increase the utility of the SSBN by incorporating conventional capabilities in addition to the traditional mission of strategic nuclear deterrence. By fielding a multi-role SSBNx, excess capacity and capability of other sea-based platforms can be eliminated, thus reducing fiscal pressures on the shipbuilding budget.

Recommendations

The *Nuclear Posture Review* (NPR) (2010) made a significant decision to retire the nuclear-armed sea-launched cruise missile (TLAM-N).[39] This is a misguided and mission-limiting decision that should be reversed. While American non-strategic (tactical) nuclear weapon inventories have been dramatically reduced since the end of the Cold War, complete reliance on the B-61 gravity bomb mated to retiring F-15s, F-16s, and yet to be fielded F-35s, as well as air-launched cruise missile (ALCM)-armed B-52s, exposes the "tactical" nuclear force to unwarranted risk.

The NPR describes the TLAM-N as a redundant weapon within the stockpile due to the ability to engage targets with strategic ICBMs and SLBMs. Ballistic missile overflight problems negate this logic as rapidly emerging targets—a justification for the employment of ICBMs—would most likely require overflight of Russian and/or Chinese airspace in direct contradiction to US nuclear operations doctrine.[40] Use of no-overflight SLBMs is equally problematic due to much higher yields of the Trident II D-5 warhead and multiple reentry vehicles. Additionally, unlike TLAM-Ns, warheads mated to either ICBMs or SLBMs would count against New START limits as deployed strategic weapons.

The few forward deployed tactical nuclear weapons supporting NATO, combined with the small stockpile maintained in the United States, substantially reduces the flexibility, capability, and credibility of extended regional deterrence. Re-equipping the current SSGN force with nuclear-armed TLAM-Ns and designing a TLAM-N capability into the SSBNx will allow the United States to maintain forward-deployed tactical nuclear weapons, ensuring extended deterrence to allies and flexible attack options for the president. The lower yield submarine-deployed TLAM-N provides a unique capability of a stealthily, responsive, and secure nuclear weapon to target rapidly emerging, dynamic, and potentially non-deterrable threats throughout the world.

In its 2001 NPR, the Bush administration called for a new strategic triad consisting of offensive strike capabilities, active and passive strategic defenses, and a new responsive American defense, development, and procurement infrastructure.[41] To deter and, if necessary, defeat emerging threats and adversaries in the twenty-first century, the 2001 NPR surmised that the offensive-strike leg of the triad can no longer rely solely on nuclear-armed forces, but must also leverage precise conventional global strike options.[42] To this end, the US Navy has repeatedly requested funding to research the Enhanced Effectiveness Initiative and the Trident II (D-5) reentry vehicle to significantly improve targeting accuracy, thus making conventionally armed SLBMs a viable option.[43] While Congress persists in denying direct funding for the conventional trident modification (CTM), the weapon systems manufacturer, Lockheed Martin, initiated low-level research and testing of the new reentry vehicle in 2002. Lockheed's test flights demonstrated improved accuracy, terminal warhead maneuverability, and the ability to slow atmospheric reentry in order to control impact angles and conditions.[44]

New global threats will continue to emerge that require a prompt strike capability where no forward-deployed forces exist or are insufficient to address the threat.[45] With these niche targets, the long-range precision-strike capability of conventional ballistic missiles (CBM) may prove the lowest risk and fastest option for strategic decision makers as global intelligence operations continue to shorten the strike decision cycle.[46] The use of conventional SLBMs for precision global strike (PGS) engagements further negates overflight concerns posed by North American-based conventional ICBMs.[47] CBM-armed Ohio-class and future SSBNx submarines will allow stealthy forward deployment of PGS weapons without the excessive and vulnerable footprint required by heavy bombers or dual-use fighters. Persistent projection of SSBNs to unstable regions will give decision makers a long-range, precise, rapidly employable, day or night conventional-strike option capable of penetrating nearly any airspace while avoiding overflight of countries capable of detecting and tracking ballistic missiles.[48]

Conclusion

As long as large inventories of nuclear weapons exist in adversary states, the risk and potential cost of global zero is too great to accept. Continued nuclear inventory reductions will eventually force the Pentagon to consider significant trade-offs in the number and type of weapons maintained by the United States.[49] Additionally, deep cuts to America's nuclear arsenal may force a shift from traditional flexible response, minimum-level-of-force use doctrine to a less credible counter-value, counter-population targeting process. Drastic reductions of delivery platforms of all types will increasingly restrict nuclear forces' flexibility, resiliency, and survivability. Unsubstantiated faith-based, minimum-deterrence nuclear force structuring will inflict unintended consequences on national objectives, capabilities, and strategies as well as impact the credibility of American extended deterrence.

The modern ballistic missile submarine, coupled with the latest Trident missile system, provides the United States a highly survivable, secure, mobile, accurate, and timely delivery system capable of directly engaging any target anywhere on the earth's surface. Faced with the reality of contracting defense budgets and declining warhead inventories, use of the SSBN will play a critical role in America's nuclear deterrence and prompt global-strike capabilities for the indefinite future. Developing and fielding a Trident-based CBM and redeploying the nuclear-armed TLAM will allow US forces to draw down strategic nuclear inventories, maintain a credible nuclear and non-nuclear second-strike capability, and guarantee America's extended deterrence commitments.

While the world's oceans are not transparent they are not completely opaque. Discovering, tracking, and monitoring submarines will continue to be problematic for the United States and its adversaries. However, technological advances coupled with smaller fleets will reduce the SSBNs' anonymity as it carries out its deterrence and strike missions. For this reason, while playing a critical role, the SSBN is only a portion of America's strategic triad.

Optimizing the strengths of the SSBN's delivery system by arming it with nuclear-armed cruise missiles and conventionally armed long-range strike ballistic missiles will diversify the platform further enabling flexible-response options and allowing it to expand its role from its current singular purpose. To justify the disproportional construction expense the new SSBN must be a multi-role platform allowing reductions and cuts to less critical current and future programs.

Notes

1 Barack H. Obama, public address, Hradcany Square, Prague, Czech Republic, April 5, 2009.
2 US Department of Defense, *Nuclear Posture Review Report* (Washington, DC: Office of the Secretary of Defense, April 2010), 20.
3 Barack H. Obama, public address, Pariser Platz, Brandenburg Gate, Berlin, Germany, June 19, 2013.
4 RIA Novosti, "Russia Skeptical over Obama's New Nuclear Reduction Proposal," June 19, 2013, http://en.ria.ru/russia/20130619/181755868/Russia-Skeptical-Over-Obamas-New-Nuclear-Reduction-Proposal.html. In response to Obama's Berlin comments, Russian Deputy Prime Minister Dmitry Rogozin harshly criticized the proposed reductions as "unrealistic," and in light of continued US ABM efforts "[t]o show the lack of understanding of this [by proposing further nuclear cuts]—means either openly lying, bluffing and deceiving, or demonstrating a deep lack of professionalism."
5 US Department of State, *New START Treaty Aggregate Numbers of Strategic Offensive Arms* (Washington, DC: Bureau of Arms Control, Verification and Compliance, October 1, 2013). According to the New START treaty required biannual data exchange, as of October 1, 2013, the US had 1,688 nuclear warheads and gravity bombs deployed on a mix of ICBMs and SLBMs, and unarmed heavy bombers. A total of 1,015 deployed and non-deployed ICBM and SLBM launchers and deployed and non-deployed heavy bombers were maintained by the US.
6 Rear Admiral Terry Benedict, Director of Strategic Systems Programs, statement before the Subcommittee on Strategic Forces of the Senate Armed Services Committee FY2014 Strategic Systems, April 17, 2013, 2.
7 Amy F. Woolf, *U.S. Strategic Nuclear Forces: Background, Developments, and Issues* (Washington, DC: Congressional Research Service, October 22, 2013), 8. Each Trident II is capable of carrying up to eight MIRVs depending on type of warhead loaded. Either the W-76 or heavier W-88 are able to be loaded on the Trident II SLBM. Rear Admiral Terry Benedict, statement before the Subcommittee on Strategic Forces, 16.
8 Michael J. Dobbs, "The Incredible Shrinking SSBN(X)," *US Naval Institute Proceedings*, 138, 6 (June 2012): 1.
9 Woolf, *U.S. Strategic Nuclear Forces*, 19.
10 Robert S. Norris and Hans M. Kristensen, "U.S. Nuclear Forces, 2006," *Bulletin of the Atomic Scientists*, 62, 1 (January 2006): 68–71. According to Rear Adm. Charles B. Young, USN Strategic Systems program director, the increased numbers of Ohio-class SSBNs and Trident II D-5 "enhances system accuracy, payload, and hard-target capability, thus improving [US] available responses to existing and emerging Pacific theater threats."
11 Ronald O'Rourke, *Navy Ohio Replacement (SSBN[X]) Ballistic Missile Submarine Program: Background and Issues for Congress* (Washington, DC: Congressional Research Service, March 14, 2013): 2. The 18 original Ohio-class SSBNs were fielded with 30-year service lives.
12 Benedict, statement before the Subcommittee on Strategic Forces, 4. Original Trident II D-5 service life was 25 years. It was extended to 50 years, double the service life of any previous sea-based deterrent system.
13 Christopher J. Castelli, "Navy Confronts $80 Billion Cost of New Ballistic Missile Submarines," *Inside the Navy*, Public Articles vol. 22. No. 48, December 7, 2009, http://insidedefense.com/200912071825134/Inside-Defense-General/Public-Articles/navy-confronts-80-billion-cost-of-new-ballistic-missile-submarines.html.
14 O'Rourke, *Navy Ohio Replacement*, 10–12.

15 Michael Fabey, "US Navy Defends Boomer Submarine Replacement Plans," *Aviation Week's DTI*, September 28, 2012, www.military.com/daily-news/2012/09/28/us-navy-defends-boomer-submarine-replacement-plans.html.

16 Association of the United States Navy, "Ohio Class Replacement Submarine Program (SSBN(X))," *2012 Fact Sheet*, www.ausn.org/Portals/0/pdfs/fact-sheets/Ohio%20 Class%20Replacement%20Submarine%20Program%20(SSBN(X))%202012%20Fact %20Sheet.pdf.

17 O'Rourke, *Navy Ohio Replacement*, 12.

18 Kris Osborn, "Admiral: Navy Must Shrink Submarine Development Costs," *DoD Buzz Online Defense and Acquisition Journal*, September 27, 2013, www.dodbuzz. com/2013/09/27/admiral-navy-must-shrink-submarine-development-costs/.

19 Rick Rozoff, "Prompt Global Strike: World Military Superiority Without Nuclear Weapons," *Global Research*, April 11, 2010, www.globalresearch.ca/america-s-imperial-design-prompt-global-strike-world-military-superiority-without-nuclear-weapons.

20 Richard Smoke, *National Security and the Nuclear Dilemma: An Introduction to the American Experience in the Cold War* (New York: McGraw Hill, Inc. 1993), 91.

21 Ben McGrath, "Countering the Pentagon's 'Pivot to Asia': China and Russia Hold War Games in Sea of Japan," *Global Research*, July 16, 2013, www.globalresearch. ca/countering-the-us-pivot-to-asia-china-and-russia-hold-war-games-in-sea-of-japan/ 5342881.

22 Kenneth N. Waltz, "Nuclear Myths and Political Realities," *The American Political Science Review* 84, 3 (September 1990): 738.

23 Paul K. Davis, *Studying First-Strike Stability with Knowledge-Based Models of Human Decision-making* (Santa Monica, CA: Rand Corp, 1989), 15.

24 Thomas C. Schelling, *Arms and Influence* (New Haven, CT: Yale University Press, 2008), 246.

25 Lawrence Freedman, *Deterrence* (Malden, MA: Polity Press, 2004), 36–37.

26 Bryan McGrath, "SSBN(X) as Great White Whale," *Information Dissemination*, September 19, 2013, www.informationdissemination.net/2013/09/ssbnx-as-great-white-whale.html.

27 Congressional Budget Office, *An Analysis of the Navy's Fiscal Year 2014 Shipbuilding Plan* (Washington, DC: Congressional Budget Office, October 2013), 14. This amount is normalized to 2013 dollar values and includes $19.3 billion in new ship construction, $1.1 billion in nuclear power plant refueling, and $900 million in other annual related costs.

28 Congressional Budget Office, *An Analysis of the Navy's Fiscal Year 2014 Shipbuilding Plan*, 12. The Navy's 2014 30-year shipbuilding goals are to increase the total battle force fleet from 285 ships at the end of 2013 to 306 ships. From 1984 to 2013 the Navy received approximately $14 billion (2013 dollars) annually for shipbuilding. The Navy breaks down 30-year cost plans into near-, mid-, and long-term, each 10 years in length with increasing variation and uncertainty in the later decades.

29 Congressional Budget Office, *An Analysis of the Navy's Fiscal Year 2014 Shipbuilding Plan*, 24.

30 Benedict, statement before the Subcommittee on Strategic Forces, 6. Sydney J. Freedberg Jr., "Sen. McCain Slams $2.5B Carrier Cost Increase; Navy Struggles To Fund SSBN-X, Destroyers," Breaking Defense, May 8, 2013, http://breakingdefense. com/2013/05/sen-mccain-slams-2-5b-carrier-cost-increase-navy-struggles-to-fund-ssbn-x-destroyers/.

31 Congressional Budget Office, *Long-Term Implications of the 2014 Future Years Defense Program* (Washington, DC: Congressional Budget Office, November 2013), 14.

32 Congressional Budget Office, *Long-Term Implications of the 2014 Future Years Defense Program*, 33.

33 James M. Acton, "Managing Vulnerability, Second Strike: Is the U.S. Nuclear Arsenal Outmoded?" *Foreign Affairs* (March/April 2010), www.foreignaffairs.com/articles/65993/jan-lodal-james-m-acton-hans-m-kristensen-matthew-mckinzie-and-i/second-strike.

34 William Rosenau, *Special Operations Forces and Elusive Enemy Ground Targets: Lessons from Vietnam and the Persian Gulf War* (Santa Monica, CA: RAND, 2001), 33. Coalition forces resorted to loitering over the expected deployment area with the aim of destroying launchers after the missiles had been fired.

35 Woolf, *U.S. Strategic Nuclear Forces*, 8. Under New START limits the United States will deploy 1,090 SLBMs in 14 SSBNs. With two boats undergoing maintenance and half of the remaining 12 on normal patrol approximately 467 warheads are available for attack ($1,090/14 \times 6 = \sim 467$).

36 Robert D. Critchlow, *Nuclear Command and Control: Current Programs and Issues* (Washington, DC: Congressional Research Service, May 3, 2006), 19. Permissive action links (PALs) employ physical and electronic combination locking systems that prevent unauthorized access to nuclear weapons. Unlock codes are only able to be released when launch orders are issued by the president.

37 US Department of Defense, *JP 3–12: Doctrine for Joint Nuclear Operations* (Washington, DC: Pentagon, March 15, 2005), II–1.

38 Adam Ciralsky, "Will It Fly?" *Vanity Fair*, September 16, 2013, www.vanityfair.com/politics/2013/09/joint-strike-fighter-lockheed-martin. For example, the most expensive weapon system in American history, the Air Force's F-35 Joint Strike Fighter, is estimated to cost $1.5 trillion over the program's life. The original contract estimated the cost at $233 billion for 2,852 airframes. According to the GAO, per plane cost has doubled from $81 million to $161 million. Continued cost overruns may lead to pressure to cut the number of airframes produced leading to a program death spiral. Political engineering of the production process by including manufacturing in 46 states insulates congressional oversight and constrains program criticism.

39 US Department of Defense, *Nuclear Posture Review Report*, 28.

40 US Department of Defense, *JP 3–12*, II–11.

41 Donald H. Rumsfeld, Secretary of Defense, *Nuclear Posture Review Report Forward*, January 2002, www.defense.gov/news/jan2002/d20020109npr.pdf.

42 US Department of Defense, *Nuclear Posture Review [Excerpts]*, January 8, 2002, www.stanford.edu/class/polisci211z/2.6/NPR2001leaked.pdf.

43 Amy F. Woolf, *Conventional Warheads for Long-Range Ballistic Missiles: Background and Issues for Congress* (Washington, DC: Congressional Research Service, January 26, 2009), 7.

44 Woolf, *Conventional Warheads*, 8.

45 Denshaw Mistry, Austin Long, and Bruce M. Sugden, "Going Nowhere Fast: Assessing Concerns about Long-Range Conventional Ballistic Missiles," *International Security* 34, 4 (2010): 169. Rapidly emerging PGS targeting must have accurate and actionable intelligence to support employment of CBMs against the short time windowed, pop-up threats envisioned by CBM proponents. This intelligence must be precise, reliable, timely, and comprehensive in order to justify the employment of long-range strategic conventional missiles. Forward-deployed airborne and ground-based gathering assets will be required to produce actionable targeting information, however, given the limited size and nature of these forces they may not have the strike capability to adequately engage emerging targets, or by doing so will compromise their continued collecting viability.

46 Mistry, Long, and Sugden, "Going Nowhere Fast," 171. Opponents of CBM development argue that extended lead times required for high-fidelity CBM targeting intelligence would allow other conventional forces such as manned stealth bombers, conventional cruise missiles, unmanned aerial vehicles, or special operations forces to deploy in order to engage emerging targets. Additionally, near-term targeting must

have accurate and actionable intelligence to support employment of CBMs against the short time windowed, pop-up threats envisioned by CBM proponents. This intelligence must be precise, reliable, timely, and comprehensive in order to justify the employment of long-range strategic conventional missiles.

47 US Department of Defense, *JP 3–12*, II–11.

48 M. Elaine Bunn and Vincent A. Manzo, "Conventional Prompt Global Strike: Strategic Asset or Unusable Liability?" *National Defense University: Strategic Forum*, SF no. 263, February 2011, http://csis.org/files/media/csis/pubs/110201_manzo_sf_263.pdf, 2.

49 Jeff Richardson, "Shifting from a Nuclear Triad to a Nuclear Dyad," *Bulletin of Atomic Scientists*, 65, 5 (September/October 2009): 1. Eventually the triad as currently constructed will no longer be viable as critical mass and economy of force is insufficient to sustain the capability. Re-posturing as a dyad or myad will require flexible and tailorable delivery methods.

Bibliography

Acton, James M., "Managing Vulnerability, Second Strike: Is the U.S. Nuclear Arsenal Outmoded?" *Foreign Affairs* (March/April 2010), www.foreignaffairs.com/articles/65993/jan-lodal-james-m-acton-hans-m-kristensen-matthew-mckinzie-and-i/second-strike.

Association of the United States Navy, "Ohio Class Replacement Submarine Program (SSBN(X))," *2012 Fact Sheet*, www.ausn.org/Portals/0/pdfs/fact-sheets/Ohio%20Class%20Replacement%20Submarine%20Program%20(SSBN(X))%202012%20Fact%20Sheet.pdf.

Benedict, Rear Admiral Terry, Director of Strategic Systems Programs, statement before the Subcommittee on Strategic Forces of the Senate Armed Services Committee FY2014 Strategic Systems, April 17, 2013.

Bunn, M. Elaine, and Manzo, Vincent A., "Conventional Prompt Global Strike: Strategic Asset or Unusable Liability?" *National Defense University: Strategic Forum*, SF no. 263, February 2011, http://csis.org/files/media/csis/pubs/110201_manzo_sf_263.pdf.

Castelli, Christopher J., "Navy Confronts $80 Billion Cost of New Ballistic Missile Submarines," *Inside the Navy*, Public Articles vol. 22. No. 48, December 7, 2009, http://insidedefense.com/200912071825134/Inside-Defense-General/Public-Articles/navy-confronts-80-billion-cost-of-new-ballistic-missile-submarines.html.

Ciralsky, Adam, "Will It Fly?" *Vanity Fair*, September 16, 2013, www.vanityfair.com/politics/2013/09/joint-strike-fighter-lockheed-martin.

Critchlow, Robert D., *Nuclear Command and Control: Current Programs and Issues* (Washington, DC: Congressional Research Service, May 3, 2006).

Congressional Budget Office, *An Analysis of the Navy's Fiscal Year 2014 Shipbuilding Plan* (Washington, DC: Congressional Budget Office, October 2013).

Congressional Budget Office, *Long-Term Implications of the 2014 Future Years Defense Program* (Washington, DC: Congressional Budget Office, November 2013).

Davis, Paul K., *Studying First-Strike Stability with Knowledge-Based Models of Human Decision-making* (Santa Monica, CA: Rand Corp, April 1989).

Dobbs, Michael J., "The Incredible Shrinking SSBN(X)," *US Naval Institute Proceedings*, 138, 6 (June 2012).

Fabey, Michael, "US Navy Defends Boomer Submarine Replacement Plans," *Aviation Week's DTI*, September 28, 2012, www.military.com/daily-news/2012/09/28/us-navy-defends-boomer-submarine-replacement-plans.html.

Freedberg Jr., Sydney J. "Sen. McCain Slams $2.5B Carrier Cost Increase; Navy Struggles to Fund SSBN-X, Destroyers," Breaking Defense, May 8, 2013, http://breakingdefense.com/2013/05/sen-mccain-slams-2-5b-carrier-cost-increase-navy-struggles-to-fund-ssbn-x-destroyers/.

Freedman, Lawrence, *Deterrence* (Malden, MA: Polity Press, 2004).

McGrath, Ben, "Countering the Pentagon's 'Pivot to Asia': China and Russia Hold War Games in Sea of Japan," *Global Research*, July 16, 2013, www.globalresearch.ca/countering-the-us-pivot-to-asia-china-and-russia-hold-war-games-in-sea-of-japan/5342881.

McGrath, Bryan, "SSBN(X) as Great White Whale," *Information Dissemination*, September 19, 2013, www.informationdissemination.net/2013/09/ssbnx-as-great-white-whale.html.

Mistry, Denshaw, Long, Austin, and Sugden, Bruce M., "Going Nowhere Fast: Assessing Concerns about Long-Range Conventional Ballistic Missiles," *International Security* 34, 4 (2010).

Norris, Robert S., and Kristensen, Hans M., "U.S. Nuclear Forces, 2006," *Bulletin of the Atomic Scientists*, 62, 1 (January 2006).

Obama, Barack H., public address, Hradcany Square, Prague, Czech Republic, April 5, 2009.

Obama, Barack H., public address, Pariser Platz, Brandenburg Gate, Berlin, Germany, June 19, 2013.

O'Rourke, Ronald, *Navy Ohio Replacement (SSBN[X]) Ballistic Missile Submarine Program: Background and Issues for Congress* (Washington, DC: Congressional Research Service, March 14, 2013).

Osborn, Kris, "Admiral: Navy Must Shrink Submarine Development Costs," *DoD Buzz Online Defense and Acquisition Journal*, September 27, 2013, www.dodbuzz.com/2013/09/27/admiral-navy-must-shrink-submarine-development-costs/.

RIA Novosti, "Russia Skeptical over Obama's New Nuclear Reduction Proposal," June 19, 2013, http://en.ria.ru/russia/20130619/181755868/Russia-Skeptical-Over-Obamas-New-Nuclear-Reduction-Proposal.html.

Richardson, Jeff, "Shifting from a Nuclear Triad to a Nuclear Dyad," *Bulletin of Atomic Scientists*, 65, 5 (September/October 2009).

Rosenau, William, *Special Operations Forces and Elusive Enemy Ground Targets: Lessons from Vietnam and the Persian Gulf War* (Santa Monica, CA: RAND, 2001).

Rozoff, Rick, "Prompt Global Strike: World Military Superiority Without Nuclear Weapons," *Global Research*, April 11, 2010, www.globalresearch.ca/america-s-imperial-design-prompt-global-strike-world-military-superiority-without-nuclear-weapons.

Rumsfeld, Donald H., Secretary of Defense, *Nuclear Posture Review Report Forward*, January 2002, www.defense.gov/news/jan2002/d20020109npr.pdf.

Schelling, Thomas C., *Arms and Influence* (New Haven, CT: Yale University Press, 2008).

Smoke, Richard, *National Security and the Nuclear Dilemma: An Introduction to the American Experience in the Cold War* (New York: McGraw Hill, Inc. 1993), 91.

US Department of Defense, *JP 3–12: Doctrine for Joint Nuclear Operations* (Washington, DC: Pentagon, March 15, 2005).

US Department of Defense, *Nuclear Posture Review [Excerpts]*, January 8, 2002, www.stanford.edu/class/polisci211z/2.6/NPR2001leaked.pdf.

US Department of Defense, *Nuclear Posture Review Report* (Washington, DC: Office of the Secretary of Defense, April 2010).

US Department of State, *New START Treaty Aggregate Numbers of Strategic Offensive Arms* (Washington, DC: Bureau of Arms Control, Verification and Compliance, October 1, 2013).

Waltz, Kenneth N., "Nuclear Myths and Political Realities," *The American Political Science Review* 84, 3 (September 1990).

Woolf, Amy F., *Conventional Warheads for Long-Range Ballistic Missiles: Background and Issues for Congress* (Washington, DC: Congressional Research Service, January 26, 2009).

Woolf, Amy F., *U.S. Strategic Nuclear Forces: Background, Developments, and Issues* (Washington, DC: Congressional Research Service, October 22, 2013).

9 Missile defenses and strategic nuclear arms control

Technology and policy challenges

Stephen J. Cimbala and Adam B. Lowther

Background

When the idea of a "missile defense" was initially proposed in the United States by the Cook Board (1945), the idea was well ahead of its time.[1] As the Cold War took shape and the United States and Soviet Union developed intermediate-range ballistic missiles (IRBMs) and then, by 1958, intercontinental ballistic missiles (ICBMs), American political leaders came to see ballistic missile defense as critical to the nation's security. In fact, in 1958 NSC 5802 called for making an anti-ICBM weapon system of the highest priority.[2] Between 1958 and the present the United States Army and Air Force both sought to develop effective missile defense systems. The Nike and Sentinel systems were two efforts to defeat incoming missiles that were impacted by the signing of the Anti-Ballistic Missile Treaty (1972–2002), which curtailed the employment of such systems. Perhaps the best known ballistic missile defense effort was President Reagan's Strategic Defense Initiative (SDI), better known as "Star Wars."[3]

With the abrogation of the ABM treaty in 2002 by President George W. Bush, the United States began a renewed push to develop an effective anti-ballistic missile (ABM) system that would destroy one or a small number of ballistic missiles. Unlike SDI the systems under development over the past 10–15 years were not intended to defeat a large-scale nuclear attack from Russia. However, even these limited ambitions have caused Russia to protest the development and employment of ballistic missile defense (BMD) systems in Europe, for example.[4] In part, Russian President Vladimir Putin's challenge to the United States' effort to field limited BMD systems stems from the dramatic reduction in the size of nuclear arsenals over the past two decades. Where these systems could simply be overwhelmed during the Cold War, that may not be true in the near future. This concern has led Russia to fear for the stability of its own nuclear arsenal.

To the present, missile defenses continue to pose policy challenges to issues of US–Russia and NATO–Russian security policy, but technical challenges as well. For opponents of BMD, such systems serve to destabilize Russo–American deterrence stability. In order to address this concern, as well as others, at the 2010 Lisbon summit, NATO and Russian leaders agreed to seek common ground on European missile defenses. However, the return of Vladimir Putin to the

Russian presidency in 2012 marked a new assertiveness in Russian foreign and defense policy, including arms control policy. Russia's occupation and later annexation of Crimea in March 2014 caused the US and NATO to suspend, at least temporarily, military-to-military cooperation with Russia, including discussions about European missile defenses.[5] Prospects for agreement on missile defenses, or on further reductions in US and Russian operationally deployed strategic nuclear weapons in a post-New START agreement, appeared dismal in a post-Ukrainian crisis world. Given Russia's continued support for Ukrainian separatists in Eastern Ukraine, the prospects for a return to amicable relations between the United States do not look good.[6]

In the discussion that follows, we first consider some of the political and military background pertinent to the relationship between Russian and American strategic nuclear arms limitations and missile defense. Second, we perform analyses for several cases of candidate "New START-minus" agreements allegedly under study by the Obama administration, including the possible implications of missile defenses for deterrence stability under post-New START reductions. Third, we draw conclusions about how ambitious the US and Russia can be in reducing strategic nuclear forces, not only in terms of their own security and defense requirements, but also with respect to the need for involvement of other nuclear weapons states.

Nuclear arms reduction and missile defenses

President Obama has reportedly tasked the Pentagon to develop planning scenarios for further reductions in American strategic nuclear forces. These scenarios include three options for further cuts in the numbers of operationally deployed strategic nuclear weapons below New START levels: 1,100, 800, or 400 weapons.[7] The range of options provides for small, medium, and large departures from New START limits. The more ambitious among these options will also require cooperation, not only between Russia and the US, but also among other nuclear weapons states. Whereas, for example, one might imagine the US and Russia reaching agreement on a limit of 1,100 or 1,000 operationally deployed strategic nuclear weapons without third or "nth" party participation, the political baggage for more drastic limitations would be a hard sell within the American and Russian national security establishments—unless, or until, other nuclear weapons states were brought into the agreement. The departure from a two-sided to a multi-sided negotiating forum for nuclear arms reductions presents both political and military challenges to governments, especially for their defense planners and arms control negotiators.

Evaluation of the political or military value of missile defenses in current and prospective policy terms requires that we acknowledge new possibilities and new dangers.[8] Compared to the Cold War, the US and Russia now have smaller numbers of operationally deployed strategic nuclear weapons. In addition, anti-missile defense technologies are of interest not only to the United States and potentially Russia, but also to other states who feel threatened by the spread of ballistic missiles outside Europe. For example, Japan, although its government would prefer to neither join the ranks of nuclear weapons states nor to enter into a regional nuclear arms race, is nevertheless very interested in anti-missile defenses. Japan is already cooperating with the US in developing and deploying theater missile defenses for its state territory and contiguous waters. This stance is not unreasonable from Japan's perspective, considering its proximity to North Korea, China, and other Asian nuclear powers. Missile defenses might provide for a country like Japan or South Korea an alternative "deterrent by denial" instead of a nuclear deterrent by threat of unacceptable second-strike retaliation. Anti-missile defenses could also serve as an insurance policy against accidental launches or unauthorized rogue attacks. On the other hand, missile defenses have also complicated the Russo–American relationship with respect to the eventual prospects for nuclear arms control and disarmament. President George W. Bush's decision to withdraw from the ABM Treaty announced in 2001 did not, at first, draw returning fire from the government of then-President Vladimir Putin. To the contrary, in 2002 the United States and Russia concluded the Strategic Offensive Reductions Treaty (SORT) that called for the two states to reduce their numbers of operationally deployed intercontinental weapons to within a range of 1,700–2,200 each by 2012. SORT was, of course, later superseded by New START, but SORT was an intriguing way station. Unlikely bedfellows, from the standpoint of political ideology, Bush and Putin nevertheless accomplished significant nuclear reductions with SORT compared to previous levels. They did so despite Russia's clear policy statements then and subsequently that its strategic nuclear deterrent was the military backbone of its international security and great power status.[9]

During the second terms of Presidents Bush and Putin, on the other hand, the political winds shifted and Russia used the diplomatic demarche over the Bush plan to deploy elements of a US global missile defense system in Poland and the Czech Republic. Russia's objections to the Bush European missile defense plan were as much political as they were military. Russia disliked the presence of US missile defenses so close to its borders and in a former Soviet satellite that Russia regards as part of its sphere of special interest. The years 2007 and 2008 were also times of jockeying for power and position within the Kremlin as the arrangements for succession to President Putin were being worked out. Putin's administration took a hard line against American missile defenses in Europe until the departure of the George W. Bush administration and the arrival of the Obama administration along with its "reset" policy. As Pavel Podvig has noted:

> As it turns out, missile defense is a very personal subject for the Russian president, who spoke passionately about it during his recent campaign. This passion, however, serves a pragmatic political purpose: It paints a picture of Russia as under siege, which helps deflect challenges to the legitimacy of the Russian political system.[10]

The Obama reset led to the conclusion of the New START agreement on offensive force reductions and to a temporary thaw in US–Russia and Russia–NATO relations on the issue of missile defenses.[11] But the thaw on missile defenses was temporary, and animosity over this issue returned in 2011–2012 as the Obama missile defense plan for Europe became clearer in its implications and as American and Russian presidential elections loomed larger.[12]

Then-US Secretary of Defense Chuck Hagel announced in 2013 that the Pentagon planned to cancel plans for the fourth phase of the European Phased Adaptive Approach (EPAA), regarded as the phase most objectionable to Russia as a potential neutralizer of Russia's nuclear deterrent. Neither President Putin nor his military leadership was mollified by this decision. Russia continued to demand either a change in the American plan or a Russian level of involvement and participation in designing the European BMD system that satisfies its nervous military leaders and politicians as to NATO intentions and capabilities. Russian leaders have indicated that if they are dissatisfied with respect to European missile defenses Russia will decline further cooperation in offensive nuclear arms reductions and possibly deploy missiles capable of launching nonstrategic nuclear weapons closer to Russia's borders with NATO.[13] A presentation by the Russian General Staff to a conference in Moscow in the spring of 2012 summarized the differences between Russian and NATO proposals.

As Stephen J. Blank has pointed out, influential Russian policymakers and military analysts have regarded past US–Russia dialogue on strategic nuclear arms control as a net "positive" for several reasons. First, it helps commit the United States to an arms control paradigm of mutually assured destruction (MAD) or assured retaliation based on offensive forces. Second, it projects the global impression of US–Russia nuclear parity regardless of the ups and downs of Russia's military modernization. Third, the impression of nuclear-strategic parity with the United States has spillover diplomatic benefits that support Russia's self-portrait for international audiences.[14] That portrait emphasizes Russia's status as a major power in the emerging multipolar international system that will eventually displace a unipolar American dominance of the post-Cold War years. Although it might seem contradictory according to some interpretations of international relations theory, in this case the second point supports the third. The appearance of nuclear-strategic parity supports Russia's perceived quest for a multipolar international system in which the US is less influential and Russia is more influential.

A related technical issue in a two-sided NATO–Russia deployment of advanced anti-missile and air defenses is the problem of defense suppression. In order to contribute to deterrence by denial, defenses would have to be survivable against preemptive attack by defense suppression forces.[15] These forces could be based (at least theoretically) in a variety of places, including on land, at sea, airborne, or located in space—depending on the state of weapons technology and launchers. In a two-sided deterrence competition with respect to strategic nuclear forces, each side will estimate the survivability of its offensive forces, strategic anti-missile and air defenses, and defense suppression forces.[16] With present technology, defense suppression missions might be carried out by ASAT (anti-satellite) weapons based terrestrially or airborne; by kinetic or cyber-attacks on the command, control, communication, and intelligence (C3I) systems supporting defenses; or by submarine-launched ballistic missiles (SLBM) or stealthy cruise missiles ahead of a later and larger attack.

If either side's defenses were perceived to be vulnerable to prompt defense suppression, a situation of mutually reinforcing fears of anti-defensive first strikes might lead to mistaken or deliberate strikes against the other side's defenses—or against its defense-suppression weapons, arguing that those weapons constitute a standing threat to defenses that are designed to protect one's own values, not to harm others. As Secretary of Defense Ashton B. Carter has noted,

> A BMD deployment is itself a prime target, and the system is clearly useless if it can easily be destroyed. The BMD need not be absolutely survivable, but the offense must pay a high enough price to destroy the defense that such a tactic is unattractive. The defense can of course defend itself, but attack on the defense remains for most deployment schemes the most effective tactic for the offense and hence the weakest link in the defense.[17]

Thus, a relationship of deterrence or dissuasion between two powers with strategic nuclear forces, defenses, and defense suppression forces might yield multiple operational and strategic approaches.

Regardless of the military and technical obstacles to NATO–Russia cooperation on missile defenses and nuclear arms reductions, political factors may be even more important. The policy statements of Russian leaders continue to speak of NATO as a danger to national security.[18] Russia is especially sensitive to NATO's reach into former Soviet, and now extended Russian, security space, where Russia claims a privileged interest.[19] These sensitivities to NATO visibility in post-Soviet space bordering or near to Russia extend to any plans for NATO land-based interceptors, radars, or other components of a European missile defense plan. As Jacob W. Kipp has noted, the distinction between Russian "reform of the armed forces" and "military reform" is closely related to the issue of future war as Russian military forecasters see it:

> On the one hand, reform of the Armed Forces refers to the transformation of the military forces belonging to the Russian Ministry of Defense and involves both downsizing the force and transforming it into a force that will meet the needs and requirements of Russia in the post-Cold War era. Military reform, on the other hand, is a more all-embracing process which encompasses all the military and paramilitary formations of the Russian state and addresses the core political, economic and social questions attached to raising, sustaining, training, arming, deploying, and employing the military as an element of Russian national power.[20]

Therefore, in the minds of some risk-averse Russian military planners, missile defense nullification technologies might constitute a necessary part of their deterrent, despite US claims that present BMD technologies are only directed toward regional threats such as those posed by Iran and North Korea. Russia has also responded with offsetting or balancing moves, including plans for offensive weapons with BMD countermeasures and improvements in Russian anti-missile and air defense systems already deployed.[21]

NATO–Russia cooperation on missile defense is a necessary condition for their improved collaboration on nuclear nonproliferation. Although Russian and American perspectives on the prevention of nuclear-weapons spread are not identical, they are potentially convergent on some important issues. Russia does not want to encourage nuclear-weapons spread, in general, but takes a selective approach to dealing with miscreant potential or actual proliferators. The United States, on the other hand, is more likely to oppose categorically the entry of any new states into the nuclear club and insists on reversing the North Korean membership in the club of nuclear weapons states.

These differences in perspective are not necessarily insurmountable obstacles to Russo-American cooperation on nuclear nonproliferation. US–Russia disagreements are likely to be more about tactics than about the seriousness of the threat posed by a nuclear Iran or by other Middle Eastern states reacting to an apparent Iranian nuclear weapons capability. Here the missile defense issue intersects with the nonproliferation concerns of both the United States and Russia. The United States sees the European missile defense system as contributing to nonproliferation by discouraging the spread of nuclear weapons without requiring aggressive counter-proliferation measures—such as the bombing of nuclear weapons complexes and nuclear infrastructure, or the imposition of regime change by military intervention. Russia fears that a NATO missile defense system "good enough" to deter or deflect an attack from Iran or other regional nuclear powers could grow into a larger system capable of nullifying Russia's deterrent.

This three-way entanglement among offensive nuclear arms reductions, antimissile defenses, and nonproliferation posed challenges to US–Russian and Russian–NATO security cooperation during President Obama's second term. How steep is this mountain? The next section discusses the parameters of alternative post-New START regimes and their implications.

Analysis and methodology

Measuring the problem

Nuclear arms control is an aspect of military strategy and national security policy, not a thing in itself. US and Russian decisions about nuclear arms reductions also have implications for the other states in the international system—especially for current or aspiring nuclear weapons states. On one hand, the gap between American and Russian capabilities and those of everyone else helps to impose some predictability and discipline on international practices related to arms control and nonproliferation. On the other hand, the continuing reliance by the United States and Russia on nuclear weapons and nuclear deterrence encourages other nuclear weapons states to move cautiously on disarmament. It also advertises the putative value of nuclear weapons for deterrence, defense, and diplomacy.

Could Russia and the United States, given favorable political conditions, reduce their numbers of operationally deployed strategic nuclear weapons below New START levels and still fulfill their national security objectives? The obvious answer to this question is maybe. However, the question "how far" is complicated. The step from the New START upper limit of 1,550 deployed warheads to some 1,000 is an incremental one that would presumably involve no major changes in roles, missions, or force structure. Below that level, to a limit of 800 or 400 deployed weapons, difficult trade-offs may ensure for military planners and for proponents of further accomplishments in nuclear arms control and disarmament.

We examine in more detail the implications of US-Russia strategic nuclear force reductions to various levels in the analysis that follows.[22] Notional force structures for the period 2018–2020 are posited for the two states and those force structures are subjected to nuclear force exchange modeling.[23] Each state is assigned a balanced triad of strategic nuclear forces deployed on ICBMs, SLBMs, and heavy bombers. The performances of each Russian and American force for each level of deployment are analyzed under each of four operational conditions: (1) forces are on generated alert and launched on warning of attack (GEN-LOW); (2) forces are on generated alert and ride out the attack before retaliating; (3) forces are on day-to-day alert and are launched on warning; and (4) forces are on day-to-day alert and ride out the attack.

For each simulation at benchmark maximum deployment levels of 1,550, 1,000, or 500 strategic nuclear weapons for each state, an alternative scenario is postulated with missile defenses added into the equation for both states. This step poses considerable challenges to the investigator, since no one really knows how well strategic anti-missile weapons will perform against prospective attackers. For heuristic purposes, we assigned each state a combination of anti-missile and anti-air defenses capable of a range of attrition against attacking offenses: Phase I defenses successfully intercept or otherwise deflect at least 20 percent of opposed second-strike retaliating warheads; Phase II defenses intercept or otherwise deflect at least 40 percent; Phase III, at least 60 percent; and Phase IV, at least 80 percent.

Data analysis and findings

Figures 9.1 through 9.6 summarize the forces in the analysis and the outcomes for each of the nuclear force exchanges. Figures 9.1 through 9.3 show the numbers of retaliating warheads for maximum deployments of 1,550, 1,000, and 500 warheads, without defenses; Figures 9.4 through 9.6 add anti-missile and anti-air defenses (combined) into the equation using the model previously described.

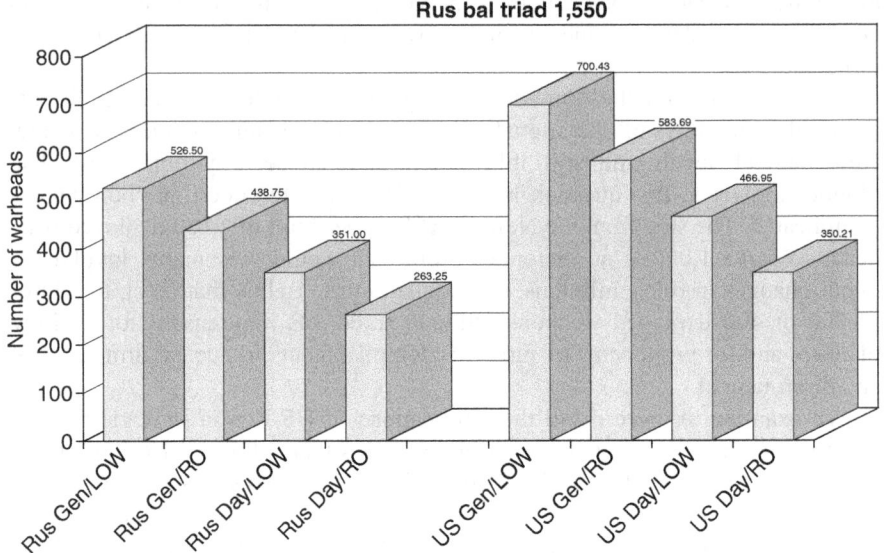

Figure 9.1 US–Russia surviving and retaliating warheads 1,550 deployment limit

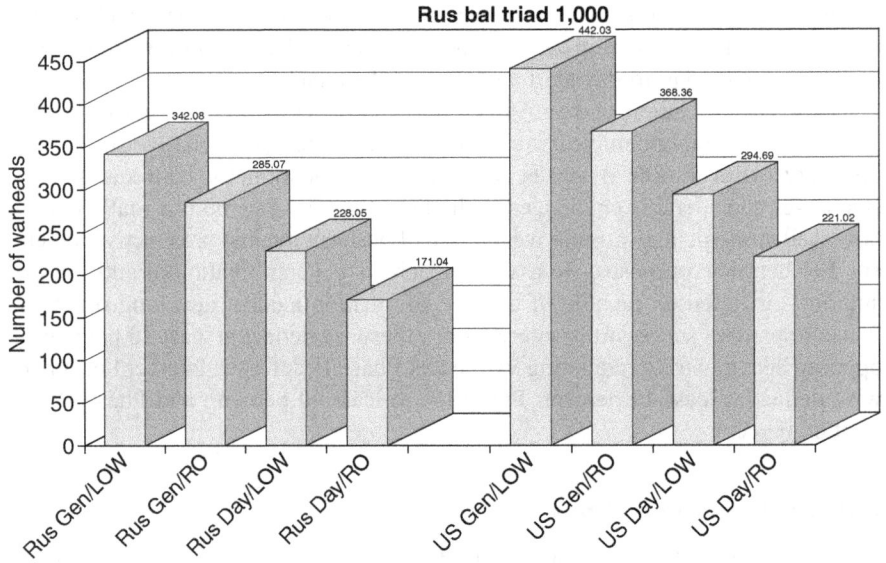

Figure 9.2 US–Russia surviving and retaliating warheads 1,000 deployment limit

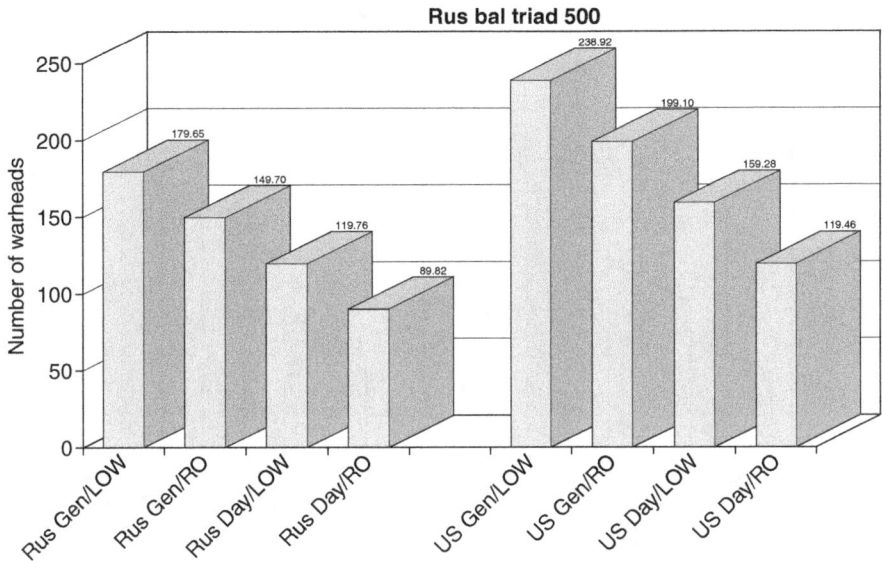

Figure 9.3 US–Russia surviving and retaliating warheads 500 deployment limit

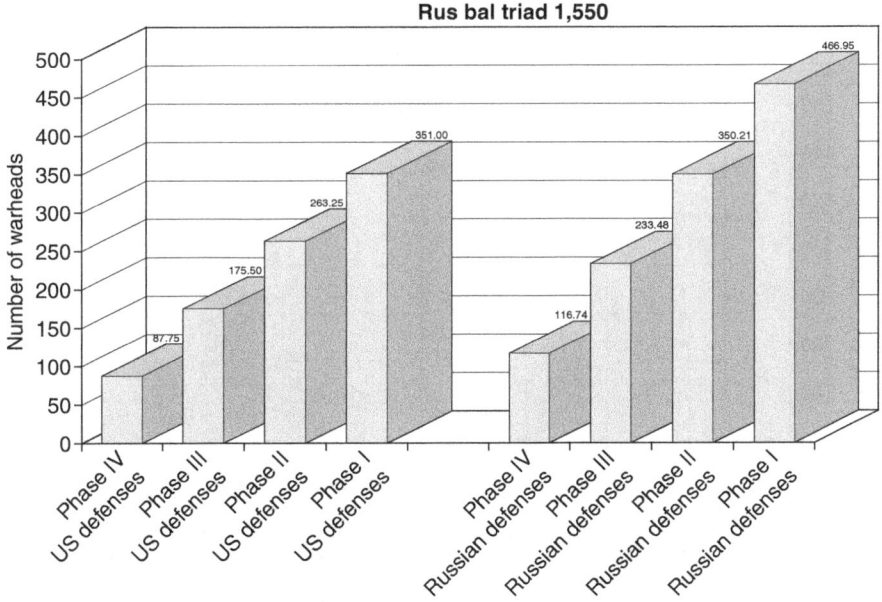

Figure 9.4 US–Russia surviving and retaliating warheads vs. defenses 1,550 deployment limit

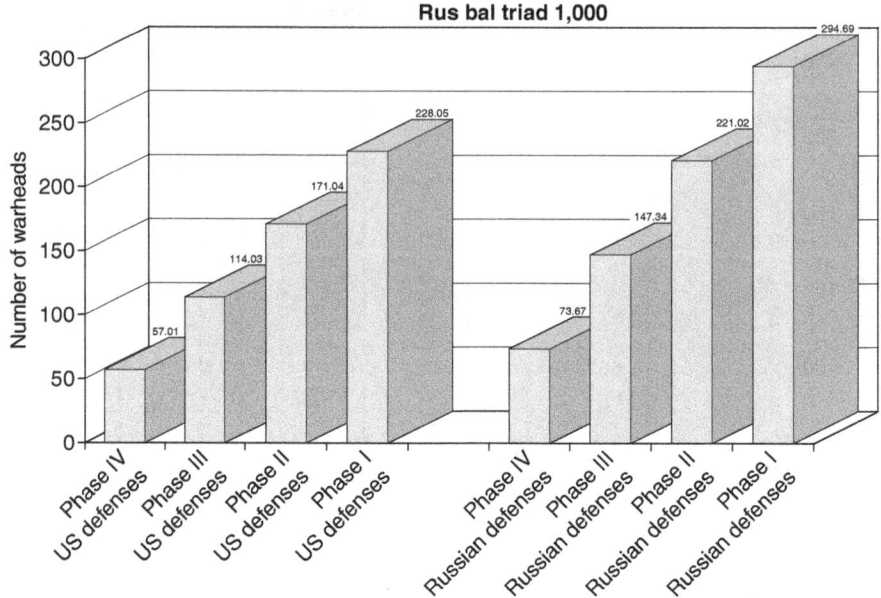

Figure 9.5 US–Russia surviving and retaliating warheads vs. defenses 1,000 deployment limit

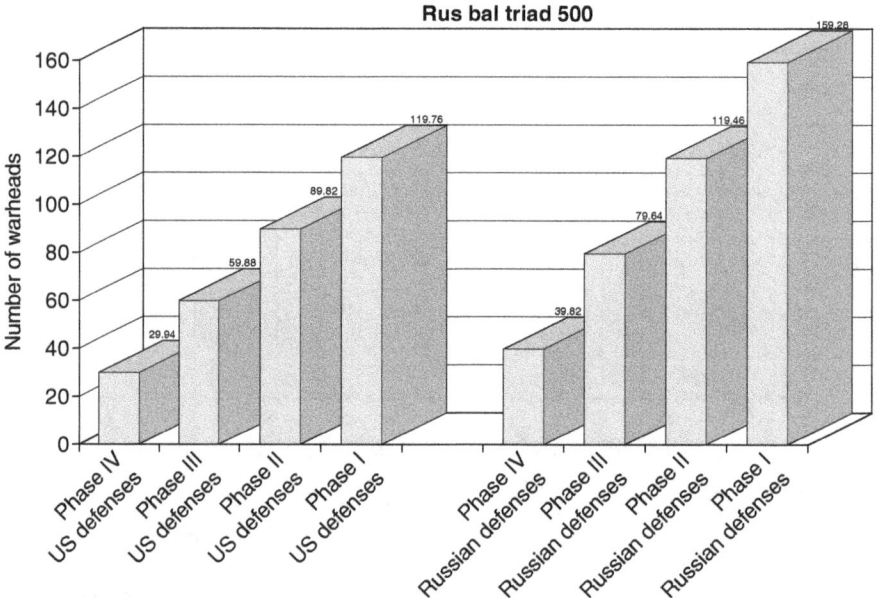

Figure 9.6 US–Russia surviving and retaliating warheads vs. defenses 500 deployment limit

If these are the relevant numbers and the models accurately predict future capability and behavior, what inferences do they suggest? First, both Russia and the United States can fulfill their deterrent and defense missions at deployment levels below New START-agreed figures—in the ideal circumstances of the model. Even the 500-weapon limit for the two states includes a considerable amount of retaliatory destruction, especially if weapons are concentrated against cities or other "soft" targets. Second, force structures do matter. The mix of land- and sea-based missiles and bombers deployed by either state can contribute to crisis and deterrence stability or detract from it. In particular, when survivability depends upon launch on warning, the potential for nuclear crisis instability is increased. For Russia, this makes it imperative that its sea-based nuclear deter-rent be rebooted with the construction of a new class of ballistic missile sub-marines and with a reliable new SLBM—as called for in past and present plans.

Third, as operationally deployed strategic nuclear weapons are reduced from 1,550 to 500, the options for nuclear target planners will be progressively restricted. A deployed force at or below 500 weapons invites an *almost exclusive* focus on counter-city or counter-value targeting. Target plans emphasizing the killing of non-combatants instead of the destruction of opposing forces might be repugnant on ethical grounds. A possible alternative to counter-city targeting is to emphasize the targeting of defense-related and other critical infrastructure, but such a target focus would require larger numbers of weapons than used in the low end of the model. An infrastructure-emphatic targeting plan would still kill many civilians but perhaps not so deliberately as would attacks targeted against populations.

Fourth, it will require some persuading to get the United States or Russia to agree to reductions below the 1,000 operationally deployed strategic nuclear weapons limit unless the additional reductions are discussed on a multilateral basis that includes the other nuclear weapons states. The United States and Russia will have mixed motives in this regard: improving the security of their relationship and disposing of unnecessary nuclear weapons, on the one hand; but, on the other hand, maintaining their role as the dominant nuclear weapons states unless, or until, other countries have signed onto a commitment for serious and verifiable reductions of their own. Getting the major nuclear weapons states of Asia into this multilateral agreement will be crucial.

Fifth, missile defenses figure ambiguously into this mix of possibilities for Russian–American offensive nuclear force reductions. US missile defenses provide talking points for Russian politicians and defense hawks, but Russians should not deceive themselves by overselling the performances of emerging US defense technologies. For this decade, at least, the European Phased Adaptive Approach or national missile defenses deployed in the continental United States can mitigate the consequences of small nuclear attacks. But exclusive theater or strategic anti-missile defenses against larger attacks will require breakthroughs in technology development and in the affordable deployment of new weapons and new launch platforms. Doubtless there are some innovative ideas about missile defenses now incubating in research laboratories and think tanks.[24]

Nevertheless, the offense-defense arithmetic in nuclear scenarios does not favor the defender because even a few nuclear weapons can do so much damage. What is harder to calculate, however, is the psychological impact on an adversary's decision calculus as they attempt to incorporate ballistic missile defense into their decision-making process. Just what role they may play in promoting deterrence by denial is uncertain.

Conclusions

Missile defenses pose technical and policy challenges for nuclear deterrence and arms control, but those challenges are not insurmountable. The technical aspect of missile defenses is whether they can continue to improve their performance envelopes and cost-effectiveness relative to offensive ballistic missiles and bomber-delivered weapons. US military planners already recognize that current and future long-range strike platforms will be opposed by increasingly advanced integrated air defenses, such as Russia's S-300 and S-400 systems. For example, prospective adversaries of the United States in Asia, for example, may seek to acquire these systems or build their own as part of their A2/AD (anti-access, area denial) as countermeasures to the United States' Air-Sea Battle concept and strategic "pivot" toward the region.[25] In its 2012 posture statement, the US Air Force notes that "as A2/AD capabilities proliferate, our [US Air Force] fourth generation fighter and legacy bomber capability to penetrate contested airspace is increasingly challenged" and that procurement of a new penetrating bomber "is critical to maintaining our [US Air Force] long-range strike capability in the face of evolving A2/AD environments."[26] While the US Air Force did award a contract for the Long-Range Strike Bomber to Northrup Grumman in October 2015, it is worth noting that the very integrated air defense systems (IADS) that threaten American bombers can also threaten the United States' ballistic missiles—giving potential adversaries their own defensive capability.

On the other hand, the US and its allies are also working to strengthen their own regional missile defenses in Asia and elsewhere, against growing ballistic and cruise missile threats to the use of the aerospace, maritime, and cyber commons. The A2/AD picture in Asia is but one illustration of the point that future missile defenses will be challenged, not only to improve their "hardware" relative to offenses, but also to enhance their "software" for scenario expectations and flexible adaptation to unexpected contingencies.[27] As a US Army study on integrated air and missile defenses has noted:

> Adversary long-range precision attack doctrines, as demonstrated in numerous experiments and service-level wargames, have evolved from a low number of missile launches from static positions to large, complex salvoes from mobile forces. Their complex precision strikes are typically supported by advanced electronic attack; offensive cyber capabilities; terrestrial and space-based intelligence, surveillance and reconnaissance (ISR); and attacks on US space-based capabilities.[28]

With regard to strategic nuclear arms control, under ideal circumstances, Russia and the United States could possibly reduce their numbers of operationally deployed strategic nuclear weapons to 1,000 or even 500 and perhaps maintain stable deterrence based on second-strike retaliation. However, the fog and friction of war and our inability to ensure that 100 percent of the nation's nuclear weapons will reach their targets should leave the results of such ideal modeling in doubt. How far, if at all, they can descend on this scale depends partly on the levels of political trust and military cooperation between Washington and Moscow, which is at an all-time low. Mutual disarmament also depends upon the cooperation of other nuclear weapons states that may have to agree to freeze or reduce their own arsenals.[29] Thus far, China—the other large player—has refused to participate in arms control negotiations.

Missile defense technologies are considerably improved compared to their Cold War predecessors. However, missile defenses as proposed in the US Phased Adaptive Approach for Europe are not "game changers" for US–Russia strategic nuclear stability. Russian defense modernization will have more to do with the viability of its nuclear deterrent than will US and NATO missile defenses. Further, the missile defense issue should not be hijacked by ideologues or partisans in Washington or Moscow. Both political and technical cooperation between NATO and Russia are possible and, in fact, desirable—although probably delayed until Putin has departed from office.

Notes

1 John Dabrowski, *Missile Defense: The First Seventy Years* (Washington, DC: Missile Defense Agency, 2013), 3.
2 John Dabrowski, *Missile Defense: The First Seventy Years*, 7.
3 Edward Reiss, *The Strategic Defense Initiative* (Cambridge, UK: Cambridge University Press, 1992), 7–15.
4 "US to Deploy ABM Systems in Europe Despite P5+1 Deal with Iran," *RT News*, December 17, 2013, https://www.rt.com/news/hagel-shoigu-missile-defense-356/.
5 "Opinion: US Pullout on Missile Defense Talks Won't Impact Russia," *RIA Novosti*, April 3, 2014, in Johnson's Russia List 2014 #74, April 3, 2014.
6 "Russia Supplying Weapons, Troops to Ukraine Separatists: Report," Agence France-Presse, April 11, 2015, http://news.yahoo.com/russia-supplying-weapons-troops-ukraine-separatists-report-225443618.html.
7 Lawrence Korb and Alex Rothman, "Obama Plan to Reduce Nukes Is Good for Budget, Boosts Moral Authority on Global Proliferation," February 15, 2012, http://thinkprogress.org/security/2012/02/15/426332/obama-plan-to-reduce-nukes-is-good-for-budget-boosts-moral-authority-on-global-proliferation/. See also Arthur Blinov, "Obama's Anti-Nuclear Signal to Russia: The United States Suggests a Dramatic Reduction of Nuclear Warheads," *Nezavisimaya Gazeta*, February 16, 2012, in Johnson's Russia List 2012 #29, February 16, 2012.
8 Pertinent expert commentary on missile defenses as related to Russia and US arms control objectives appears in Jacob W. Kipp, "Russia's Future Arms Control Agenda and Posture," Ch. 1, 1–62; and Steven Pifer, "The Russian Arms Control Agenda after New START," Ch. 2, 63–92, both in Stephen J. Blank, ed., *Russia and the Current State of Arms Control* (Carlisle, PA: Strategic Studies Institute, US Army War College, September 2012).

9 Nikolai Sokov, "The New, 2010 Russian Military Doctrine: The Nuclear Angle," Center for Nonproliferation Studies, Monterey Institute of International Studies, February 5, 2010, http://cns.miis.edu/stories/100205_russian_nuclear_doctrine.htm. See also Vladimir Putin, "Being Strong: National Security Guarantees for Russia," *Rossiiskaya Gazeta*, February 20, 2012, http://premier.gov.ru, reprinted in Johnson's Russia List 2012 #31, February 21, 2012; and "Russia Retains Right to Play Nuclear Card–Gen-Staff Chief," www.russiatoday.com, in Johnson's Russia List 2012 #29, February 16, 2012.

10 Pavel Podvig, "Point of Distraction," Russian Strategic Nuclear Forces Blog, June 1, 2012, http://russianforces.org/blog/2012/06/point_of_distraction.shtml.

11 *Treaty between the United States of America and the Russian Federation on Measures for the Further Reduction and Limitation of Strategic Offensive Arms* (Washington, DC: US Department of State, April 8, 2010), www.state.gov/documents/organization/140035.pdf.

12 The Obama administration's European Phased Adaptive Approach to missile defense will retain and improve some technologies deployed by the George W. Bush administration, but shift emphasis to other interceptors supported by improved battle-management-command-control-communications (BMC3) systems and launch detection and tracking. See Karen Kaya, "NATO Missile Defense and the View from the Front Line," *Joint Force Quarterly* 71 (4th Quarter 2013), 84–89. See also Association of the US Army (AUSA), *US Army Integrated Air and Missile Defense Capabilities: Enabling Joint Force 2020 and Beyond* (Washington, DC: Institute of Land Warfare, AUSA, May 2014); Steven J. Whitmore and John R. Deni, *NATO Missile Defense and the European Phased Adaptive Approach: The Implications of Burden Sharing and the Underappreciated Role of the US Army* (Carlisle, PA: Strategic Studies Institute, US Army War College, October 2013); Patrick J. O'Reilly, *Ballistic Missile Defense Overview, Presented to 10th Annual Missile Defense Conference* (Washington, DC: US Department of Defense, March 26, 2012), www.mda.mil/news/downloadable_resources.html; North Atlantic Treaty Organization, *NATO Ballistic Missile Defense (BMD), Fact Sheet* (Brussels: North Atlantic Treaty Organization, May 22, 2012), www.nato.int/nato_static/assets/pdf/pdf_topics/20120520_media-backgrounder_NATO_ballistic_missile_defence_en.pdf; and The White House, Office of the Press Secretary, "Fact Sheet on US Missile Defense Policy: A 'Phased, Adaptive Approach' for Missile Defense in Europe," September 17, 2009, www.whitehouse.gov/the_press_office/FACT-SHEET-US-Missile-Defense-Policy-html. For critical assessment of US missile defense plans by expert scientists and other commentators, see Committee on an Assessment of Concepts and Systems for US Boost-Phase Missile Defense in Comparison to Other Alternatives, *Making Sense of Ballistic Missile Defense: An Assessment of Concepts and Systems for U.S. Boost-Phase Missile Defense in Comparison to Other Alternatives* (Washington, DC: National Research Council, National Academy of Sciences, National Academies Press, 2012); William J. Broad, "U.S. Missile Defense Strategy Is Flawed, Expert Panel Finds," *New York Times*, September 11, 2012, www.nytimes.com/2012/09/12/science/us-missile-defense-protections-are-called-vulnerable.html, downloaded September 13, 2012; Tom Z. Collina, "Failure to Launch: Why Did America Just Spend $30 Billion on a Missile Defense System That Doesn't Work?," *Foreign Policy*, September 12, 2012, http://foreignpolicy.com/2012/09/13/failure-to-launch/; Philip Coyle, "The Failures of Missile Defense," *The National Interest*, July 26, 2012, http://nationalinterest.org/print/commentary/the-failures-missile-defense-7248; and George N. Lewis and Theodore A. Postol, "A Flawed and Dangerous U.S. Missile Defense Plan," *Arms Control Today*, May 2010, www.armscontrol.org/act/2010_05/Lewis-Postol.

13 For additional perspective on this topic, see Daniel Wagner and Diana Stellman, "The Prospects for Missile Defense Cooperation Between NATO and Russia," *Foreign Policy Journal* (February 2011), in Johnson's Russia List 2011 #24, February 10,

2011; and Stephen J. Blank, *Arms Control and Proliferation Challenges to the Reset Policy* (Carlisle, PA: Strategic Studies Institute, US Army War College, November 2011), 32–33.

14 See Blank, *Arms Control and Proliferation Challenges to the Reset Policy*.

15 Paul K. Davis suggests that dissuasion by denial (DND) is the preferable term in "Toward Theory for Dissuasion (or Deterrence) by Denial: Using Simple Cognitive Models of the Adversary to Inform Strategy," *RAND Working Paper* (Santa Monica, CA: RAND National Security Research Division, WR-1027, January, 2014).

16 The topic of defense suppression receives more detailed treatment in Dean Wilkening, Kenneth Watman, Michael Kennedy, and Richard Darilek, *Strategic Defenses and Crisis Stability* (Santa Monica, CA: RAND Corporation, April 1989), 35–40.

17 Ashton B. Carter, "BMD Applications: Performance and Limitations," Ch. 4 in Ashton B. Carter and David N. Schwartz, eds., *Ballistic Missile Defense* (Washington, DC: Brookings Institution, 1984), 98–181, citation p. 106.

18 Russia's 2010 military doctrine refers to "dangers" as well as "threats," whereas prior editions made reference only to threats. Although "dangers" might seem less menacing than "threats" to interested readers, the dangers mentioned are concrete and specific compared to the threats, the latter of a more general nature. Listed dangers include the desire of NATO to globalize its force potential and move its military infrastructure closer to the borders of Russia. See Marcel de Haas, "Russia's Military Doctrine Development (2000–10)," Ch. 1 in Stephen J. Blank, ed., *Russia's Military Politics and Russia's 2010 Defense Doctrine* (Carlisle, PA: Strategic Studies Institute, US Army War College, March 2011), 1–61.

19 For additional historical perspective on Russian military doctrine, see Jacob W. Kipp, "Russian Military Doctrine: Past, Present, and Future," Ch. 2 in Blank, ed., *Russia's Military Politics and Russia's 2010 Defense Doctrine*, 63–151. See also "The Military Doctrine of the Russian Federation," text, www.Kremlin.ru, February 5, 2010, in Johnson's Russia List 2010 #35, February 19, 2010.

20 Jacob W. Kipp, *Forecasting Future War: Andrei Kokoshin and the Military-Political Debate in Contemporary Russia* (Ft. Leavenworth, KS: Foreign Military Studies Office, January 1999), http:www.fas.org/nuke/guide/Russia/agency/990100-kokoshin.htm.

21 Nicholas Khoo and Reuben Steff, "This Program Will Not Be a Threat to Them: Ballistic Missile Defense and US Relations with Russia and China," *Defense and Security Analysis* 1 (March 2014): 17–28.

22 Force structures are the author's. For expert estimates, see Joseph Cirincione, "Strategic Turn: New US and Russian Views on Nuclear Weapons," *New America Foundation*, June 29, 2011, http://newamerica.net/publications/policy/strategic_turn; and Pavel Podvig, "New START Treaty in Numbers," Russian Strategic Nuclear Forces, April 9, 2010, http://russianforces.org/blog/2010/03/new_start_treaty_in_numbers.shtml.

23 Grateful acknowledgment is made to Dr. James J. Tritten for the use of a model originally developed by him and modified by the author. Dr. Tritten is not responsible for any of the analysis or arguments here.

24 For example, a study by Global Zero discusses the possibility of missile defenses augmented by passive defenses (such as hardening and sheltering) and advanced US conventional missions against regional adversaries such as Iran or North Korea. See James Cartwright, *US Nuclear Policy Commission Report: Modernizing U.S. Nuclear Strategy, Force Structure and Posture* (Washington, DC: Global Zero, 2012).

25 For expert analysis, see Jeremiah Gertler, *US Air Force Bomber Sustainment and Modernization: Background and Issues for Congress* (Washington, DC: Congressional Research Service, June 4, 2014), R43049.

26 Department of the Air Force, *United States Air Force Posture Statement* (Washington, DC: United States Air Force, 2012), 15–16, cited in Gertler, *US Air Force Bomber Sustainment and Modernization*, 5–6. For definitions of A2/AD capabilities, see p. 6, note 18.

27 Rebecca Slayton, *Arguments that Count: Physics, Computing and Missile Defense, 1949–2012* (Cambridge, MA: The MIT Press, 2013), 216–219.
28 Association of the US Army (AUSA), *US Army Integrated Air and Missile Defense Capabilities: Enabling Joint Force 2020 and Beyond* (Washington, DC: Institute of Land Warfare, AUSA, May 2014), 12.
29 On the need for a multilateral approach to nuclear arms reductions, see Association of the US Army (AUSA), *US Army Integrated Air and Missile Defense Capabilities: Enabling Joint Force 2020 and Beyond*, 3–4.

Bibliography

Association of the US Army (AUSA), *US Army Integrated Air and Missile Defense Capabilities: Enabling Joint Force 2020 and Beyond* (Washington, DC: Institute of Land Warfare, AUSA, May 2014).

Blank, Stephen J., *Arms Control and Proliferation Challenges to the Reset Policy* (Carlisle, PA: Strategic Studies Institute, US Army War College, November 2011).

Blank, Stephen J., ed., *Russia and the Current State of Arms Control* (Carlisle, PA: Strategic Studies Institute, US Army War College, September 2012).

Blinov, Arthur, "Obama's Anti-Nuclear Signal to Russia: The United States Suggests a Dramatic Reduction of Nuclear Warheads," *Nezavisimaya Gazeta*, February 16, 2012, in Johnson's Russia List 2012 #29, February 16, 2012.

Broad, William J., "U.S. Missile Defense Strategy Is Flawed, Expert Panel Finds," *New York Times*, September 11, 2012, www.nytimes.com/2012/09/12/science/us-missile-defense-protections-are-called-vulnerable.html, downloaded September 13, 2012.

Carter, Ashton B., "BMD Applications: Performance and Limitations," Ch. 4 in Ashton B. Carter and David N. Schwartz, eds., *Ballistic Missile Defense* (Washington, DC: Brookings Institution, 1984).

Cartwright, James, *US Nuclear Policy Commission Report: Modernizing U.S. Nuclear Strategy, Force Structure and Posture* (Washington, DC: Global Zero, 2012).

Cirincione, Joseph, "Strategic Turn: New US and Russian Views on Nuclear Weapons," *New America Foundation*, June 29, 2011, http://newamerica.net/publications/policy/strategic_turn.

Collina, Tom Z., "Failure to Launch: Why did America Just Spend $30 Billion on a Missile Defense System That Doesn't Work?," *Foreign Policy*, September 12, 2012, http://foreignpolicy.com/2012/09/13/failure-to-launch/.

Committee on an Assessment of Concepts and Systems for US Boost-Phase Missile Defense in Comparison to Other Alternatives, *Making Sense of Ballistic Missile Defense: An Assessment of Concepts and Systems for U.S. Boost-Phase Missile Defense in Comparison to Other Alternatives* (Washington, DC: National Research Council, National Academy of Sciences, National Academies Press, 2012).

Coyle, Philip, "The Failures of Missile Defense," *The National Interest*, July 26, 2012, http://nationalinterest.org/print/commentary/the-failures-missile-defense-7248.

Dabrowski, John, *Missile Defense: The First Seventy Years* (Washington, DC: Missile Defense Agency, 2013).

Davis, Paul K., "Toward Theory for Dissuasion (or Deterrence) by Denial: Using Simple Cognitive Models of the Adversary to Inform Strategy," *RAND Working Paper* (Santa Monica, CA: RAND National Security Research Division, WR-1027, January, 2014).

de Haas, Marcel, "Russia's Military Doctrine Development (2000–10)," Ch. 1 in Stephen J. Blank, ed., *Russia's Military Politics and Russia's 2010 Defense Doctrine* (Carlisle, PA: Strategic Studies Institute, US Army War College, March 2011).

Department of the Air Force, *United States Air Force Posture Statement* (Washington, DC: United States Air Force, 2012), cited in Gertler, *US Air Force Bomber Sustainment and Modernization.*

Gertler, Jeremiah, *US Air Force Bomber Sustainment and Modernization: Background and Issues for Congress* (Washington, DC: Congressional Research Service, June 4, 2014), R43049.

Kaya, Karen, "NATO Missile Defense and the View from the Front Line," *Joint Force Quarterly* 71 (4th Quarter 2013).

Khoo, Nicholas, and Steff, Reuben, "This Program Will Not Be a Threat to Them: Ballistic Missile Defense and US Relations with Russia and China," *Defense and Security Analysis* 1 (March 2014).

Kipp, Jacob W., *Forecasting Future War: Andrei Kokoshin and the Military–Political Debate in Contemporary Russia* (Ft. Leavenworth, KS: Foreign Military Studies Office, January 1999), www.fas.org/nuke/guide/Russia/agency/990100-kokoshin.htm.

Kipp, Jacob W., "Russian Military Doctrine: Past, Present, and Future," Ch. 2 in Stephen J. Blank, ed., *Russia's Military Politics and Russia's 2010 Defense Doctrine* (Carlisle, PA: Strategic Studies Institute, US Army War College, March 2011).

Kipp, Jacob W., "Russia's Future Arms Control Agenda and Posture," in Blank, ed., *Russia and the Current State of Arms Control.*

Korb, Lawrence, and Rothman, Alex, "Obama Plan to Reduce Nukes Is Good for Budget, Boosts Moral Authority on Global Proliferation," February 15, 2012, http://thinkprogress.org/security/2012/02/15/426332/obama-plan-to-reduce-nukes-is-good-for-budget-boosts-moral-authority-on-global-proliferation/.

Lewis, George N., and Postol, Theodore A., "A Flawed and Dangerous U.S. Missile Defense Plan," *Arms Control Today*, May 2010, www.armscontrol.org/act/2010_05/Lewis-Postol.

"The Military Doctrine of the Russian Federation," www.Kremlin.ru, February 5, 2010, in Johnson's Russia List 2010 #35, February 19, 2010.

North Atlantic Treaty Organization, *NATO Ballistic Missile Defense (BMD), Fact Sheet* (Brussels: North Atlantic Treaty Organization, May 22, 2012), www.nato.int/nato_static/assets/pdf/pdf_topics/20120520_media-backgrounder_NATO_ballistic_missile_defence_en.pdf.

O'Reilly, Patrick J., *Ballistic Missile Defense Overview, Presented to 10th Annual Missile Defense Conference* (Washington, DC: US Department of Defense, March 26, 2012), www.mda.mil/news/downloadable_resources.html.

"Opinion: US Pullout on Missile Defense Talks Won't Impact Russia," *RIA Novosti*, April 3, 2014, in Johnson's Russia List 2014 #74, April 3, 2014.

Pifer, Steven, "The Russian Arms Control Agenda after New START," in Stephen J. Blank, ed., *Russia and the Current State of Arms Control* (Carlisle, PA: Strategic Studies Institute, US Army War College, September 2012).

Podvig, Pavel, "New START Treaty in Numbers," Russian Strategic Nuclear Forces, April 9, 2010, http://russianforces.org/blog/2010/03/new_start_treaty_in_numbers.shtml.

Podvig, Pavel, "Point of Distraction," Russian Strategic Nuclear Forces Blog, June 1, 2012, http://russianforces.org/blog/2012/06/point_of_distraction.shtml.

Putin, Vladimir, "Being Strong: National Security Guarantees for Russia," *Rossiiskaya Gazeta*, February 20, 2012, http://premier.gov.ru, reprinted in Johnson's Russia List 2012 #31, February 21, 2012.

Slayton, Rebecca, *Arguments that Count: Physics, Computing and Missile Defense, 1949–2012* (Cambridge, MA: The MIT Press, 2013).

Reiss, Edward, *The Strategic Defense Initiative* (Cambridge, UK: Cambridge University Press, 1992).

"Russia Retains Right to Play Nuclear Card–Gen-Staff Chief," www.russiatoday.com, in Johnson's Russia List 2012 #29, February 16, 2012.

"Russia Supplying Weapons, Troops to Ukraine Separatists: Report," Agence France-Presse, April 11, 2015, http://news.yahoo.com/russia-supplying-weapons-troops-ukraine-separatists-report-225443618.html.

Sokov, Nikolai, "The New, 2010 Russian Military Doctrine: The Nuclear Angle," Center for Nonproliferation Studies, Monterey Institute of International Studies, February 5, 2010, http://cns.miis.edu/stories/100205_russian_nuclear_doctrine.htm.

Treaty between the United States of America and the Russian Federation on Measures for the Further Reduction and Limitation of Strategic Offensive Arms (Washington, DC: US Department of State, April 8, 2010), www.state.gov/documents/organization/140035.pdf.

"US to Deploy ABM Systems in Europe Despite P5+1 Deal with Iran," *RT News*, December 17, 2013, https://www.rt.com/news/hagel-shoigu-missile-defense-356/.

Wagner, Daniel, and Stellman, Diana, "The Prospects for Missile Defense Cooperation Between NATO and Russia," *Foreign Policy Journal* (February 2011) in Johnson's Russia List 2011 #24, February 10, 2011.

White House, Office of the Press Secretary, "Fact Sheet on US Missile Defense Policy: A 'Phased, Adaptive Approach' for Missile Defense in Europe," September 17, 2009, www.whitehouse.gov/the_press_office/FACT-SHEET-US-Missile-Defense-Policy-html.

Whitmore, Steven J., and Deni, John R., *NATO Missile Defense and the European Phased Adaptive Approach: The Implications of Burden Sharing and the Underappreciated Role of the US Army* (Carlisle, PA: Strategic Studies Institute, US Army War College, October 2013).

Wilkening, Dean, Watman, Kenneth, Kennedy, Michael, and Darilek, Richard, *Strategic Defenses and Crisis Stability* (Santa Monica, CA: RAND Corporation, April 1989).

10 Nuclear modernization and the nonproliferation treaty

Compliance or compromise?

Shelley Bischoff Kavlick

Introduction

The most terrifying invention known to man is the nuclear weapon. Society's fear of these awesome devices is evident in our unwillingness to use these weapons over the past seven decades. During the Cold War, the stakes were too high for the United States or Soviet Union to even approach their use. Despite the end of the Cold War, nuclear proliferation has occurred as countries pursue nuclear programs for a variety of motives that include energy, security, or influence. In order to prevent the spread of nuclear weapons and ultimately achieve complete disarmament, the United States government and other countries joined together in the late 1960s to create the Treaty on the Non-Proliferation of Nuclear Weapons (NPT). President Barack Obama proclaimed America's renewed commitment to the NPT during his Prague speech in April 2009. International support for the NPT is hopeful at best and does not address the realities of living in a world where nations with nuclear weapons have no intention of relinquishing them. In spite of the aspirational goals established in the NPT, the evidence suggests that the day and age of nuclear technology is here to stay, which is why Nuclear Weapons States (NWS) must responsibly secure and maintain their respective nuclear arsenals.

Treaty on the nonproliferation of nuclear weapons

The NPT entered into force on March 5, 1970, under the agreement of three main pillars: prevent the spread of nuclear weapons and weapons technology (Articles I and II); foster the peaceful uses of nuclear energy (Articles III, IV, and V); and further the goal of disarmament (Article VI).[1] The NPT has been ratified by 189 nations. (North Korea withdrew when it declared its nuclear program and India, Pakistan, and Israel have abstained.)[2] The International Atomic Energy Agency (IAEA) is responsible for monitoring compliance by member states with nonproliferation safeguards. While the IAEA has demonstrated success in identifying proliferation violations, the utopian goals of the NPT are proving difficult to achieve in a technically oriented and politically charged geostrategic security environment where the IAEA lacks the authority and capability to effectively safeguard against proliferation. NPT inspections, for example, require consent to visit declared sites as dictated by member states. While stronger inspection protocols were added in 1997 to the original 1972 Safeguards Agreement, after the discovery of Iraq's clandestine nuclear weapons program, only 128 of 189 NPT member states have adopted these requirements.[3]

While the NPT is an internationally recognized treaty, it relies upon the willingness of states to comply with its requirements. Article X of the NPT provides the option for member states to withdraw at any time. Furthermore, original provisions call for a NPT Review Conference 25 years after the date of NPT enforcement to determine whether to indefinitely extend the treaty. Member states met as prescribed in 1995 and decided to extend NPT Review Conferences in increments of five years. According to the 2013 edition of the *2010 NPT Action Plan Monitoring Report*, NPT member states, and specifically the NWS, show little interest or ability to support the comprehensive 64-point action plan, which provides a more detailed path to denuclearization, leaving much work remaining to fully implement a number of provisions by the next review conference in 2015.[4] Perhaps the idea of "global zero," or a world free of nuclear weapons, is not incentive enough to realize the NPT's full aspirations. The notion that the NPT equitably applies to Non-Nuclear Weapons States (NNWS) is a matter of perspective and trust between NNWS and NWS. The IAEA safeguards and inspections apply only to NNWS. There remains a perception of a double standard between NNWS and NWS. NPT prohibitions and obligations fall on NNWS to forego acquisition of nuclear weapons while NWS are expected to negotiate disarmament without threat of consequences by failure to do so.[5] The grand bargaining that is at the heart of the NPT is the dilemma for NNWS to refrain from developing nuclear weapons while NWS continue to rely on their nuclear arsenals. The original NWS or the P5 countries that are permanent members of the United Nations (UN) Security Council (China, France, Russia, United Kingdom, and United States) plus Israel, India, Pakistan, and North Korea actively maintain and modernize their nuclear arsenals and with no plans to abolish weapons. The intent of NWS to remain armed with nuclear weapons is evident by their active modernization efforts and reluctance to ratify the Comprehensive Nuclear Test Ban Treaty (CTBT).

Comprehensive Nuclear Test Ban Treaty

In an effort to ban nuclear testing to promote nuclear nonproliferation, the UN General Assembly adopted the CTBT nearly two decades after the NPT in 1996. The CTBT specifically bans nuclear weapon tests and other nuclear-related explosions. The treaty will enter into force once signed and ratified by the United States and other member states with either nuclear power or research reactors. As of June 2013, of the 44 required states, 41 have signed and 36 ratified the CTBT, of which the United States has signed but not ratified.[6] The United States has not conducted nuclear testing since 1992 while other NWS have tested as recently as 2013—Russia in 1990, India and Pakistan in 1998, and North Korea in 2013.[7] Advocates of the CTBT assert by the treaty entering into force positive steps will be taken to fulfill the intent of Article VI of the NPT. Perceived inequities between NNWS and NWS would be mitigated as the P5 states agree to forego the right to test and develop sophisticated new warheads.[8] While the current administration is intent on ratifying the CTBT, Congress remains divided on the merits of this treaty.

The National Academy of Sciences' (NAS) March 2012 report, *The Comprehensive Nuclear-Test-Ban Treaty: Technical Issues for the United States*, provides some considerations associated with the CTBT. According to the NAS, "[T]echnical capabilities for maintaining the stockpile absent nuclear explosion testing are better now than anticipated."[9] Modern research and development efforts may possibly eliminate reliance on explosive tests to ensure reliability of legacy nuclear weapons. However, like criticism of the NPT, "there is currently no mechanism that would enable Congress to assess whether CTBT safeguards were being fulfilled after entry into force."[10] The United States, as other NWS, is inherently responsible to the international community to maintain safe and reliable nuclear arsenals. Diplomatic pressure to reduce nuclear stockpiles is forcing US policymakers to address the relevance of nuclear weapons with respect to the current and future national security environment. American policy on nuclear test and modernization can be understood by examining the *Nuclear Posture Review* (NPR).

Nuclear Posture Review

The NPR currently plays an important role in establishing the long-range policy of the current administration. The 2010 NPR expands upon President Obama's address to the international community in Prague (2009) where he communicated his greater vision for nuclear weapon disarmament while advocating for the national security interests of the United States and its allies. More specifically, the NPR states that "fundamental changes in the international security environment ... enable us to fulfill those objectives ... with reduced reliance on nuclear weapons ... without jeopardizing our traditional deterrence and reassurance goals."[11] Top priority for US nuclear security is the prevention of nuclear proliferation and nuclear terrorism to be achieved through three identified objectives. First, the NPR aims to bolster the nonproliferation regime through the NPT; second, to accelerate efforts to secure vulnerable nuclear materials worldwide; and third, to pursue arms control efforts through international treaties.[12] Such treaties include the CTBT and New Strategic Arms Reduction Treaty (New START) that limits US–Russian inventories to 1,550 deployable strategic nuclear warheads.[13]

The NPR states that nonproliferation and arms reduction will be achieved by America's commitment to the ratification of international treaties and investment in the aging nuclear infrastructure. Then-Secretary of Defense Robert Gates subsequently called for $5 billion to be transferred from the Department of Defense (DoD) to the Department of Energy (DOE) for warhead life extension in support of a modernization plan for nuclear deterrence.[14] Based on language of the current NPR it is clear that the United States will proceed with support of nuclear nonproliferation efforts while seeking innovative approaches to keeping a formidable deterrent to nuclear threats. The following discussion defines the nuclear employment strategy.

Nuclear employment strategy

In accordance with Section 491 of 10 US Code, the Secretary of Defense led an interagency follow-on analysis of the 2010 NPR in 2011 known as the *Report on Nuclear Employment Strategy*. The report reiterated President Obama's directive to achieve a credible nuclear deterrent for both the US and its allies with a reduced nuclear force structure. The DoD and DOE recommend a three-pronged approach to hedge against risk to the credibility of the US nuclear stockpile. The strategy proposed three recommendations:

1   Maintain ample non-deployed nuclear weapons to offset technical failure of any weapon and delivery system by providing intra-leg options (i.e., interchangeability between triad warheads);
2   Maintain legacy weapons until confidence in life-extension programs are achieved; and
3   Ensure right-sized and ready non-deployed weapons hedge for flexibility to respond to geopolitical events that alter deployed force requirements.[15]

Investment in the US nuclear enterprise is an important theme in the NPR and follow-on nuclear employment strategy. America's nuclear weapons policy not only provides strategic direction and resource allocation guidance to the nuclear enterprise, but it also informs the international community on US nuclear deterrence priorities.

Relevance of US nuclear policy to the international community

The current administration is actively engaging with the international community on its nuclear weapons policy through various means (NPT, CTBT, New START, and NPR) for a variety of reasons. At present, the United States is the premier nuclear superpower in both force structure and capability. P5 NWS may follow the lead of the United States, or they may seek asymmetric advantage. While nuclear nonproliferation is the collective responsibility of the international community, America is looked upon to take the initiative for others to follow. The *National Security Strategy* (NSS) of 2010 declares that the United States will "pursue the goal of a world without nuclear weapons."[16] The current administration recognizes the role expected of the United States by fellow NPT member states to lead the cause against nonproliferation. The dilemma that inherently lies in total nuclear disarmament is that as long as known states and possibly non-state actors hold nuclear weapons, nuclear deterrence will remain a national security concern. The NSS goes on to assert that there is "no greater threat to the American people than weapons of mass destruction, particularly the danger posed by the pursuit of nuclear weapons by violent extremists and their proliferation to additional states."[17] For these reasons, the NSS further states that the United States will invest in modernization of the nuclear arsenal to ensure a "safe, secure and effective stockpile" without new production.[18] We can expect that the US will continue to invest in the nuclear enterprise to remain a relevant deterrent. However, how will a reduced nuclear force structure affect US nuclear deterrence? The answer is partly dependent on the nonproliferation regime and the perception of a credible extended deterrence.

The NSS validates America's commitment to nuclear nonproliferation by extending a negative security assurance to non-nuclear NPT member states not to use or threaten the use of nuclear weapons against these nations.[19] The US assures partner nations that, by agreeing not to proliferate, America's nuclear force structure will maintain a credible extended deterrent against an adversarial nuclear strike. Multilateral alliances and bilateral agreements could be jeopardized if America's declining nuclear force structure were perceived as less than credible. By communicating the intent to invest in the modernization of the nuclear enterprise, the United States may not only maximize capability with a reduced arsenal, but also mitigate a negative perception of extended deterrence credibility. Another key element to nuclear deterrence is how current NWS and emerging NWS interpret the credibility of United States nuclear policy.

Nuclear weapons stabilize the security environment by their contribution to deterrence. The NPR clearly supports this theory, stating America will support "strategic stability through an assured second strike capability."[20] Nuclear weapons remain the ultimate deterrent against total war between states with assets that can be held at risk.[21] Furthermore, nuclear weapons are an attractive solution for states facing significant security threats that cannot afford large conventional militaries. Regionally armed NWS such as Pakistan and India balance the threat of rival nuclear states with their own deterrent weapons. The rise of current and aspiring NWS like North Korea and Iran are representative of countries seeking increased international influence and to balance power with other nuclear-armed states within their respective regions. In the future, the nuclear club may grow significantly larger and unstable. An increase in NWS will shape US policy. For the American arsenal to remain credible in such a strategic environment modernizing the nuclear enterprise is required. While a declining nuclear inventory may signal good faith towards arms reduction, modernization efforts can be perceived as a silent arms race among NWS and possibly invoke proliferation. What is certain is that the post-Cold War nuclear landscape is no longer bipolar.

The global nuclear security environment is growing ever more complex and dynamic as nuclear weapons potentially become more accessible to states and non-state actors. The National Defense University Center for the Study of Weapons of Mass Destruction characterizes the world with high nuclear latency.

They suggest that, while nuclear weapons programs are the greatest threat to nuclear proliferation, other latent possibilities just below this threshold or capability ought to be considered.[22] Advanced technologies that enable states to develop nuclear programs for peaceful purposes can lead to processes that conceal threshold or virtual nuclear weapons capabilities. This movement is not limited to recognized states alone. The possibility of non-state actors and individuals proliferating nuclear technology is a dangerous reality. Consider the fallout of nuclear black-market sales by the Pakistani nuclear scientist A. Q. Kahn who trafficked nuclear weapons technology to North Korea and Iran. Given the lack of legally binding consequences of member states withdrawing from the NPT to develop indigenous nuclear weapons programs, nuclear latency is a potential prelude to proliferation should regional security become a concern for NNWS.

The phenomenon of nuclear latency and the continued threat of near-peer competitors are incentives for NWS to modernize their nuclear programs. NWS all maintain and plan to modernize their weapons. Global zero estimates that the nine nuclear-armed nations invested $104.9 billion on their combined nuclear arsenals in 2011.[23] While US policymakers have lobbied Congress for investment funds, the bureaucracy to modernize is a slow-going progress. Former Secretary of Defense Gates said,

> No one has designed a new nuclear weapon in the United States since the 1980s, and no one has built a new one since the 1990s.... [T]he United States is the only declared nuclear power that is neither modernizing its nuclear arsenal nor has the capability to produce a new nuclear warhead.[24]

Such sentiment calls into question the relevancy of the American nuclear weapons complex. Cold War-era nuclear weaponry designed for high yield performance may not be the right deterrent for next-generation nuclear threats that include both nation states and extremist non-state actors. There is also growing concern over preserving American nuclear intellectual property. As scientists and engineers with practical nuclear design and test experience from the 1980s and 1990s near retirement, the United States will lose a generation of nuclear weapons expertise.[25] This dilemma merits further examination and rationale behind whether the nuclear enterprise should maintain the status quo or modernize to mitigate risk.

Maintaining the status quo

The Cold War experience reinforced the nuclear taboo, or the assumed inhibitions felt by states in undertaking nuclear first use or even extremely dangerous brinkmanship. In 1991, the United States and Russia made a commitment to an arms reduction treaty by agreeing to START and renewed this pledge some 20 years later by ratifying the New START in 2011. Since the end of the Cold War, NWS have grown by only four nations from the original P5 countries, indicating the relatively slow pace of nuclear weapon proliferation. Furthermore, there is growing international support for nuclear disarmament. The NPT has been ratified by 189 governments and the CTBT has been ratified by 36 of 41 required nations to enter into force. According to a 2008 world public opinion poll, 77 percent of Americans are in favor of banning nuclear weapons.[26] The Obama administration envisions a world without nuclear weapons as prescribed in the NPT and is committed to ratification of the CTBT. As the global steward of nuclear nonproliferation, the United States may influence other NWS to abandon their modernization efforts by foregoing its own modernization program. Modernization of the US nuclear enterprise would be perceived as contradictory to signed treaties and possibly incite a nuclear arms race. However, history has shown that America continues to deter other countries from attack with its current nuclear arsenal.

Advancements in conventional weapons can complement strategic deterrence while maintaining the current nuclear arsenal. Technology has produced more accurate and lethal conventional weapons that do not hold the same public scrutiny for use as nuclear weapons in war. Nuclear scientists and engineers could apply their technical skills toward developing conventional weapons with better kinetic effects. The likelihood of the United States using nuclear weapons again remains to be seen. There are, however, specific types of targets—such as hardened and deeply buried—that no conventional munitions is likely to be successful in destroying—despite precision guidance.

US policy on nuclear second-strike reinforces the notion that nuclear weapons will only be used in retaliation of a deliberate nuclear attack against America or its allies. Continued development and revolutions in conventional weapons technology do not present a perceived compromise of America's commitment to international nonproliferation treaties. Modernization of conventional weapons does not counter the principles behind the NPT, but rather deemphasizes focus on the prestige of nuclear weapons. The status quo argument is feasible so long as the national security landscape remains relatively unchanged with predictable threats. However, today's threats may grow into tomorrow's fears as global security concerns are constantly evolving. In order to prepare for this uncertain future, the US nuclear arsenal and weapons complex should remain relevant not only to current but future security challenges that are hard, if not impossible, to predict.

Modernize to mitigate risk

Critics of nuclear weapons modernization may ask: Why reinvest in the US nuclear enterprise when nuclear weapons have not been used in almost 70 years? The answer to this question is relatively simple; nuclear weapons have been used every single day of every year for nearly seven decades as a strategic deterrent and a preventer of conventional conflict. In fact, nuclear weapons have been so successful that there has neither been a nuclear attack nor a great power war since 1945. There is a clear distinction between nuclear proliferation, where states compete to acquire the same nuclear capability, and nuclear deterrence, where states prevent one another from using nuclear weapons. While the NPT aims to reduce the overall number of nuclear weapons in the world, it is important to recognize that it does not prevent the use of nuclear weapons. Former Secretary of State Hillary Clinton has said that "the nuclear status quo is neither desirable nor sustainable.... [I]t gives other countries the motivation or excuse to pursue their own nuclear programs."[27] A representative from the State Department Bureau of Arms Control, Verification, and Compliance states that production (modernization included) of new nuclear weapons is not banned under international law.[28] The US position is to adhere to the intent of the CTBT while attempting to affirm the integrity of current nuclear weapons without resorting to underground nuclear explosive testing.[29] If technology could support modernization of nuclear components within a warhead without explosive testing, then modernization would not violate the intent of the CTBT. A relevant nuclear weapons program is a tool of counterproliferation only if it remains a credible deterrent to dissuade growth of NWS and transferability to NNWS.

So this leaves the question: In order to endure as a credible strategic deterrent, can the United States modernize without violating the intent of the NPT? Earlier attempts at nuclear modernization were explored under the reliable replacement warhead (RRW) program funded by Congress between 2004 and 2008 where the National Nuclear Security Administration (NNSA) researched the feasibility of replacing aging warheads in the nuclear stockpile.[30] The RRW program was intended to make nuclear warheads easier to manufacture, more environmentally friendly, possess larger margins for reliability, and eliminate underground certi-fication testing.[31] The goal of the program centered on versatility of both a common platform warhead and a modifiable nuclear yield.[32] Skeptics believed confidence in the RRW could not be proven without nuclear testing while others saw merit in the program as an opportunity to train the next generation of nuclear weapons designers. The Obama administration eventually cancelled the program in 2009 due to lack of funding. Thereafter modernization efforts for the nuclear enterprise became a function of the Life Extension Program (LEP).

The NPR states that the US will not develop new nuclear warheads or conduct nuclear testing and will use only those nuclear components from previously tested designs.[33] Warhead updates would be generated from nuclear components that are refurbished, reused, or replaced. The NPR recommends the LEP for W-76 (submarine), W-78 (ICBM), and the B-61 (airborne) warheads. US policy limitations on new design testing may call into question the integrity of LEP warheads. Former Secretary of Defense Gates expressed concern with the LEP as a long-term solution for the nuclear arsenal stating that "with every adjust-ment, we move farther away from the original design that was successfully tested when the weapon was first fielded."[34] At what point will America's inven-tory of nuclear warheads fail certification? *The Report on Nuclear Employment Strategy* calls for the DoD to maintain legacy warheads until confidence is reached for each LEP. In essence, ratified nuclear stockpile reductions may be delayed due to the progress and outcome of the LEP.

More recent developments in modernization efforts include the "3+2 Plan." The joint DoD and DOE Nuclear Weapons Council endorsed a 25-year path toward a long-term stockpile solution. The 3+2 Plan proposes warhead replacement and interoperability among three ballistic missile warheads (submarines and land-based) and two air-delivered warheads (cruise missiles and bombs).[35] Such an initiative would require design modification and possible legal or political complications. However, new technology has evolved to certify warheads inside the laboratory in lieu of field test. Modernization offers the prospects of engineering right-sized nuclear warheads with flexibility to meet today and tomorrow's security challenges without increasing the size of the nuclear arsenal. America's pool of nuclear expertise would not be at critical mass, rather modernization can train the next generation of nuclear scientists and engineers. The United States could remain New START-compliant by replacing antiquated warheads with reengineered warheads on a one-for-one basis and possibly seek further arms reductions with more capable nuclear warheads providing increased accuracy, lower yields, and reduced collateral damage. More significantly, the 3+2 Plan would give Congress more options whether or not to ratify the CTBT.

Approach to policy and recommendations

US policymakers should stay the course with America's renewed commitment to the nuclear enterprise so that the United States remains a credible strategic deterrent. The nuclear status quo is based on a legacy Cold War security strategy that lacks the flexibility and confidence to deter current and emerging security threats. A relevant twenty-first-century nuclear arsenal, a research and development program, and supporting infrastructure will signal to allies and adversaries alike that America will remain an unrivaled global nuclear leader. Modernization will also improve the adaptability of the nuclear arsenal by developing lower yield nuclear weapons with tailored options for the National Command Authority to respond to an array of security threats. Technology now permits new warhead certification in controlled laboratory conditions without violating the CTBT. Certified confidence in warhead accuracy and collateral damage would enable a leaner nuclear arsenal. The alternative viewpoint is that America's modernization efforts may incite an arms race with members of the nonproliferation regime. NWS are already actively updating their nuclear arsenals. America's modernization efforts should include diplomatic engagement with the UN Security Council to minimize international political backlash and perception of proliferation. Transparent development and strategic dialogue with the IAEA, NWS, and NNWS will communicate the United States's intent to be a global leader in nuclear weapons efficiency and arms reduction.

The United States will remain in compliance with the NPT should it pursue nuclear modernization. America would continue to lead nonproliferation efforts by investing in aging nuclear warheads to provide a relevant and credible extended deterrent for its allies while demonstrating the highest of standards by safeguarding the advanced technology. The United States can continue to assist

NNWS to foster their own peaceful nuclear energy programs. The US can advance President Obama's goal toward further nuclear disarmament by replacing antiquated warheads with reengineered warheads, thereby improving the performance and efficiency of the nuclear arsenal by enabling stockpile reductions that may inspire other NWS to follow suit. America's modernization efforts will signal to the nonproliferation regime that the US remains committed to the tenets of the NPT while maintaining a safe and responsible nuclear force.

Follow-on nuclear policy debate should examine nuclear modernization as a symbiotic relationship between nuclear and conventional weapons. When, if possible, could advancements in the precision and lethality of conventional weapons someday produce the same effects as nuclear weapons? If so, how much should the US balance investment between revolutionizing conventional weapons and modernizing the nuclear enterprise? These are difficult questions senior policymakers should consider in an approach to informing the 2014 NSS and addressing the security challenges that lie ahead.

Notes

1 International Atomic Energy Agency, Treaty on the Non-Proliferation of Nuclear Weapons, INFCIRC/140, April 22, 1970, www.iaea.org/publications/documents/infcircs/treaty-non-proliferation-nuclear-weapons.
2 Reaching Critical Will, "Nuclear Non-Proliferation Treaty," www.reachingcriticalwill.org/disarmament-fora/npt.
3 David Cortright and Raimo Vayrynen, *Towards Nuclear Zero* (New York: Routledge, 2009), 125–126.
4 Reaching Critical Will, *The NPT Action Plan Monitoring Report*, March 2013, 10, www.reachingcriticalwill.org/resources/publications-and-research/publications/5456-npt-action-plan-monitoring-reports.
5 Cortright and Vayrynen, *Towards Nuclear Zero*, 36–37.
6 Jonathan Medalia, "Comprehensive Nuclear Test Ban Treaty: Background and Current Developments," Congressional Research Service, June 10, 2013, ii, https://www.fas.org/sgp/crs/nuke/RL33548.pdf.
7 Medalia, "Comprehensive Nuclear Test Ban Treaty: Background and Current Developments."
8 Medalia, "Comprehensive Nuclear Test Ban Treaty: Background and Current Developments," 20.
9 Medalia, "Comprehensive Nuclear Test Ban Treaty: Background and Current Developments," 46.
10 Medalia, "Comprehensive Nuclear Test Ban Treaty: Background and Current Developments," 46.
11 US Department of Defense, *Nuclear Posture Review Report*, v.
12 US Department of Defense, *Nuclear Posture Review Report*, vi–vii.
13 US Department of Defense, *Nuclear Posture Review Report*, ix.
14 US Department of Defense, *Nuclear Posture Review Report*, i.
15 US Department of Defense, *Report on Nuclear Employment Strategy of the United States Specified in Section 491 of 10 U.S.C.*, June 12, 2013, 7, www.defense.gov/pubs/reporttoCongressonUSNuclearEmploymentStrategy_Section491.pdf.
16 White House, *National Security Strategy*, May 2010, 23, www.au.af.mil/au/awc/awcgate/nss/nss_may2010.pdf.
17 White House, *National Security Strategy*, 4.

18 White House, *National Security Strategy*, 23.
19 White House, *National Security Strategy*, 23.
20 US Department of Defense, *Nuclear Posture Review Report*, 20.
21 Stephen M. Younger, *The Bomb: A New History* (New York: Harper Collins, 2009), 208.
22 Paul I. Bernstein, *The Future Nuclear Landscape* (Washington, DC: National Defense University Press, 2007), 1.
23 Don't Bank on the Bomb 2013 Report, "Nuclear Forces: An Overview of Nuclear Weapons Modernization," dontbankonthebomb.com, 32, www.dontbankonthebomb.com/wp-content/uploads/2012/02/Chapter3.pdf.
24 Robert A. Monroe, "A Perfect Storm over Nuclear Weapons," *Air & Space Power Journal* 23 (Fall 2009): 25.
25 Edgar M. Vaughan, *Recapitalizing Nuclear Weapons* (Maxwell AFB, AL: AU Press, 2007), 8.
26 Don't Bank on the Bomb, "Nuclear Forces," 32.
27 Cortright and Vayrynen, *Towards Nuclear Zero*, 43.
28 Department of State senior official, e-mail, February 4, 2014.
29 US Department of State, "Maintaining the US Nuclear Weapons Stockpile in the Absence of Nuclear Explosive Testing," September 28, 2012, www.state.gov/t/avc/rls/198238.htm.
30 "Reliable Replacement Warhead," Wikipedia, http://en.wikipedia.org/wiki/Reliable_Replacement_Warhead.
31 Richard Garwin, "A Different Kind of Complex: The Future of U.S. Nuclear Weapons and the Nuclear Weapons Enterprise," *Arms Control Today* 38 (December 2008): 14.
32 Sarah J. Diehl and James C. Moltz, *Nuclear Weapons and Non-Proliferation* (Santa Barbara, CA: ABC CLIO, 2008), 78.
33 US Department of Defense, *Nuclear Posture Review Report*, xiv.
34 Garwin, "A Different Kind of Complex," 14.
35 Union of Concerned Scientists, "Making Smart Choices: The Future of the US Nuclear Weapons Complex," October 2013, 9, www.ucsusa.org/assets/documents/nwgs/nuclear-weapons-complex-report.pdf.

Bibliography

Bernstein, Paul I., *The Future Nuclear Landscape* (Washington, DC: National Defense University Press, 2007).

Cortright, David, and Vayrynen, Raimo, *Towards Nuclear Zero* (New York: Routledge, 2009).

Department of State senior official, e-mail, February 4, 2014.

Diehl, Sarah J., and Moltz, James C., *Nuclear Weapons and Non-Proliferation* (Santa Barbara, CA: ABC CLIO, 2008).

Don't Bank on the Bomb 2013 Report, "Nuclear Forces: An Overview of Nuclear Weapons Modernization," dontbankonthebomb.com, www.dontbankonthebomb.com/wp-content/uploads/2012/02/Chapter3.pdf.

Garwin, Richard, "A Different Kind of Complex: The Future of U.S. Nuclear Weapons and the Nuclear Weapons Enterprise," *Arms Control Today* 38 (December 2008).

International Atomic Energy Agency, Treaty on the Non-Proliferation of Nuclear Weapons, INFCIRC/140, April 22, 1970, www.iaea.org/publications/documents/infcircs/treaty-non-proliferation-nuclear-weapons.

Medalia, Jonathan, "Comprehensive Nuclear Test Ban Treaty: Background and Current Developments," Congressional Research Service, June 10, 2013, www.fas.org/sgp/crs/nuke/RL33548.pdf.

Monroe, Robert A., "A Perfect Storm over Nuclear Weapons," *Air & Space Power Journal* 23 (Fall 2009).

Reaching Critical Will, *The NPT Action Plan Monitoring Report*, March 2013, www.reachingcriticalwill.org/resources/publications-and-research/publications/5456-npt-action-plan-monitoring-reports.

Reaching Critical Will, "Nuclear Non-Proliferation Treaty," www.reachingcriticalwill.org/disarmament-fora/npt.

"Reliable Replacement Warhead," Wikipedia, http://en.wikipedia.org/wiki/Reliable_Replacement_Warhead.

Union of Concerned Scientists, "Making Smart Choices: The Future of the US Nuclear Weapons Complex," October 2013, www.ucsusa.org/assets/documents/nwgs/nuclear-weapons-complex-report.pdf.

US Department of Defense, *Nuclear Posture Review Report*.

US Department of Defense, *Report on Nuclear Employment Strategy of the United States Specified in Section 491 of 10 U.S.C.*, June 12, 2013, www.defense.gov/pubs/reportto-CongressonUSNuclearEmploymentStrategy_Section491.pdf.

US Department of State, "Maintaining the US Nuclear Weapons Stockpile in the Absence of Nuclear Explosive Testing," September 28, 2012, www.state.gov/t/avc/rls/198238.htm.

Vaughan, Edgar M., *Recapitalizing Nuclear Weapons* (Maxwell AFB, AL: Air University Press, 2007).

White House, *National Security Strategy*, May 2010, www.au.af.mil/au/awc/awcgate/nss/nss_may2010.pdf.

Younger, Stephen M., *The Bomb: A New History* (New York: Harper Collins, 2009).

11 Should the United States ratify the Comprehensive Nuclear Test Ban Treaty or is nuclear testing still necessary?

Karyn E. McKinney

Introduction

The Spanish conquistadors named a 60-mile stretch of desert in New Mexico the Jordana del Muerto Valley or "route of the dead man."[1] This portion of the desert had once been the deadliest and toughest part of the Camino Real, the highway that connected old Mexico to Santa Fe. Despite the lack of water, harsh temperatures, and the presence of hostile Indians, the valley was the preferred route because it was wide enough for supply wagons to traverse. However, on July 16, 1945, its remote location and arid climate made it the perfect location for the United States to conduct the first nuclear weapon test.[2]

Since that time, over 2,000 nuclear tests have been conducted all over the world. Concerns over the effects of radiation have set many against the use of nuclear weapons for any purpose to include testing. In response to the increased concern over the environmental effects of radiation, a partial ban on nuclear testing was established in 1963, the Limited Test Ban Treaty (LTBT). It prohibits all nuclear weapons tests or any other nuclear explosion in the atmosphere, in outer space, and under water.[3] The treaty does not ban underground testing so long as the test takes place within the territorial limits of the state conducting the explosion. Among the 126 states that have ratified the treaty are Russia, the United States, and the United Kingdom.

Even with the LTBT in effect, nuclear-armed countries continued to conduct nuclear testing underground. Nonetheless environmental concerns continued to plague nuclear programs. Thus, in 1996, the Comprehensive Nuclear Test Ban Treaty (CTBT) became the focus of nuclear arms control efforts when it was adopted by the United Nations General Assembly.[4] The CTBT prohibits all nuclear weapons testing including underground testing.[5] Although the treaty has been signed by 183 states and ratified by 159, it has not been put into force.[6] Annex 2 of the treaty contains a list of 44 states that must ratify the treaty before it can be entered into force (see the Appendix at the end of the chapter). The list was developed from an April 1996 edition of the International Atomic Energy's "Nuclear Power Reactors in the World," which identified those nations either conducting nuclear research, possessing nuclear reactors, or both.[7] As of June 2013, 41 of those states had signed the treaty, but only 36 had ratified it.[8]

President Clinton was the first to sign the treaty, but Senate deliberations in 1999 failed to bring about the treaty's ratification. Although the United States has not ratified the treaty, it has not conducted any nuclear weapon tests since 1992.[9]

The debate over whether or not the treaty should be ratified continues. Supporters of ratification argue that the existing weapons stockpile has been tested and can be maintained under the Stockpile Stewardship Program, the treaty is verifiable through the International Monitoring System, and ratification proves the United States is sincere about deemphasizing nuclear weapons.[10] Opponents of the treaty note that the current nuclear weapons arsenal continues to age and deteriorate despite the best efforts of nuclear scientists. Furthermore, the International Monitoring System will not keep states honest, as there are ways to evade detection. Finally, the resumption of nuclear weapons testing would reinforce the credibility of the United States's nuclear deterrent.[11]

The national security of the United States hinges on its ability to maintain a credible nuclear deterrent. The aging nuclear arsenal cannot guarantee a credible deterrent. If the United States ratifies the treaty and other outliers do not, confidence in the nuclear enterprise of the United States would be further degraded.

The CTBT: rationale for ratification by the United States

The Stockpile Stewardship Program and the Life Extension Program

The primary mission of the nuclear arsenal is to assure, dissuade, and deter.[12] In 1991, when President George H. W. Bush terminated all nuclear weapons production, it became imperative to find ways to maintain and verify the capabilities of the stockpile. The subsequent moratorium on nuclear weapons testing in 1992 created a swift change from nuclear weapons modernization programs to indefinite retention programs.[13] The Stockpile Stewardship Program (SSP) and the Life Extension Program (LEP) were established for the purpose of maintaining a safe and reliable arsenal.

The SSP is a highly specialized program for maintaining the safety and dependability of the stockpile in a time without nuclear testing or the development of new weapons systems.[14] It has three main goals. First, the program supports the nuclear deterrent of the United States with a safe, secure, and reliable stockpile while downsizing the nuclear weapons inventory. Second, it aims to preserve the competencies of scientists in the weapons laboratories utilizing a science-based approach. And third, it ensures that maintenance of the nation's nuclear deterrent is compatible with the nation's arms control.[15]

The LEP and SSP work in conjunction with repairing or replacing components of nuclear weapons to ensure readiness of the weapons should they be called upon for military action.[16] To ensure readiness, the warheads are certified annually, a process allowing them to remain in the stockpile beyond original life expectancy. Through the LEP the National Nuclear Security Administration (NNSA) has been able to recondition warheads that would have otherwise been dismantled and retired. The program maintains warheads by replacing deteriorated, non-functional components with newly manufactured ones that, to the extent possible, match the original. The LEP also provides scientists with the opportunity to assess the impact of aging on radiation hardness during the lifetime of the overall weapons system.[17]

The SSP and LEP utilize non-nuclear experiments, computer simulations, and analyses of data from previous nuclear tests to evaluate and assure the stockpile's safety and efficacy.[18] By utilizing these tools, scientists are better able to assess the performance and identify and fix problems with the weapons. The directors of the three national laboratories (Lawrence Livermore, Los Alamos, and Sandia) say they understand more now about how nuclear weapons work than they did during the era of explosive testing.[19]

In the absence of testing, the annual certification process provides a formal appraisal of the nation's stockpile of nuclear warheads and bombs to the president.[20] The annual certification is an essential tool for ensuring confidence in the nuclear enterprise. This assessment is based on a thorough evaluation of the stockpile using scientific and engineering tools to assess safety and potential performance. The final memorandum to the president is the culmination of 12 months of surveillance, computer simulations, component-level experiments, and subcritical experiments.[21] To date, there have been no safety or reliability concerns with the stockpile. Therefore, supporters of the CTBT argue that underground testing does not need to resume, because the health of the stockpile can be safely and reliably maintained through LEP and SSP.

Computer-simulated nuclear weapons testing

Several years ago, a computer simulation conducted at the Lawrence Livermore National Laboratory modeled the life cycle of a nuclear warhead from the moment it leaves storage to the moment of impact.[22] The test indicated flaws that could lead to catastrophic failure, meaning the weapon would either not produce the expected explosive yield, or it would produce nothing at all. Further investigation revealed errors in the way the weapon was handled prior to deployment. These flaws would not have been recognized without the assistance of computer simulation.

Since the early days of the Manhattan Project, nuclear scientists have relied on experimental data and simulations conducted with state-of-the-art computers. Computer modeling provides a better understanding of weapon physics, and has resolved issues related to aging and design flaws.[23] Today, "supercomputers" used by the national laboratories can replicate the physical impact of nuclear explosions with ultra-precise detail and incredible speed. These computers simulate molecular-scale reactions occurring within milliseconds.

Supercomputers perform three-dimensional interactive simulations, offering a candid view of the nuclear device's behavior and detailing what is happening at different points in time. The Advanced Simulation and Computing (ASC) program is a pillar of the SSP, and allows scientists to take advantage of the capabilities of the new generation of supercomputers.[24] Scientists can now visualize and analyze each component of a nuclear weapon as it goes through the primary explosion. This provides deeper understanding of how thermonuclear explosions occur and how materials behave at extreme temperatures.

The ASC program has upgraded its computing capability with a supercomputer that can process 20 quadrillion calculations (called floating point operations) per second.[25] The faster the computer, the more detailed the information it produces and the less uncertainty in its data output. The ability to process large packages of information enables the computer to run verification programs to cross-reference data and predict the most likely outcomes. Weapons codes are fed into the computer including the size and shape of the weapon's components, chemical makeup of materials, and the various phenomena acting inside the weapon during an explosion.[26] Supercomputing has produced detailed insight that was never possible before.

With the US moratorium on testing, the SSP, and supercomputing have filled the void of information created by the lack of testing and new weapons development. Computer simulation allows the United States to maintain confidence in its nuclear stockpile.

The international monitoring system

Two key components of the CTBT include monitoring for explosions consistent with the magnitude created by a nuclear detonation, and the ability to verify the occurrence of an explosion. Annex 1 of the CTBT identifies the protocol for monitoring and verifying nuclear explosions, and spells out the geographical coordinates for each monitoring site. Verification of nuclear weapons explosions has been a major sticking point for those opposed to the treaty. Advances in the International Monitoring System (IMS) have addressed this concern.

The CTBT calls for the establishment of 337 monitoring facilities (321 monitoring stations and 16 radionuclide laboratories) located all over the world that constantly monitor for signs of nuclear explosions. Although the CTBT has not been entered into force, 80 percent of the monitoring sites have already been activated.[27] Once established, each station must undergo a certification process to ensure it has implemented specific data and communication protocols, authentication devices, and interfaces with the Global Communications Infrastructure.[28] Additionally, each station must demonstrate operational practices consistent with IMS standards. Collected data is translated to the International Data Center at the Comprehensive Test Ban Treaty Organization (CTBTO) headquarters in Vienna. Data is subsequently shared with member states (those who have signed the treaty).

The IMS uses four technologies to detect nuclear weapons explosions. Seismic monitoring measures shockwaves in the earth, hydro-acoustic technology measures sound waves in the ocean, infrasound technology is used to detect low frequency sound waves emitted by large explosions, and radionuclide stations monitor radioactive particles and noble gases in the atmosphere.

When North Korea conducted its third nuclear weapons test in February 2013, sensors within the IMS detected seismic activity and alerted the international community that an explosion had occurred. Since naturally occurring seismic activity within North Korea is low, it is not likely the event was an earthquake. The first data were reported within one hour enabling the CTBTO to determine the magnitude, location, and depth of the test.[29]

If an IMS station detects a nuclear explosion, member states can request an on-site inspection to gather evidence that will assist in a final determination of whether or not an explosion has actually taken place. However, on-site inspections can only be utilized if the treaty is entered into force. Supporters of the CTBT are confident that while no treaty is 100 percent verifiable, the IMS will make it virtually impossible for any nation to explode a nuclear weapon without being detected.

The CTBT: rationale against ratification by the United States

Analysis of the Test Ban Treaty

Those in favor of the United States's ratification of the CTBT focus their argument on the successes of the SSP and LEP, the robust capabilities of the supercomputers, and the ability of the IMS to detect nuclear explosions. However strong their arguments may be, there are inadequacies with each of these points, as they can never surpass the knowledge gained by an actual test. In addition to the political statement a nuclear test makes, testing provides real-time data about the weapon's reliability, how factors such as temperature and delivery method impact its effectiveness, and how adjacent structures react.[30]

One of the primary concerns of all nations is the preservation of their national security. The development of nuclear weapons programs is a response to external threats and insecurity. As such, it is feasible to believe that, even today, some non-nuclear states would work to develop their own nuclear weapons program. For example, Iran has long been suspected of enriching uranium for the purpose of building a nuclear bomb and not for the sole purpose of energy, as it claims. Additionally, it is rumored that in return for funding to support Pakistan's nuclear weapons program, Saudi Arabia can claim some of those weapons at will.[31] As the face of the nuclear threats becomes even more unclear, now is not the time for the United States to commit to a legally binding ban on nuclear testing.

The CTBT is directed at banning all nuclear weapons testing, to include underground explosions. The inherent language within treaties is often a cause for debate, but the CTBT has created one of the longest in history. For starters, it is a treaty of unlimited duration, and the treaty does provide a clear definition of testing, the very thing it proposes to ban.

In paragraph one of Article I, the treaty states that "each state party undertakes not to carry out any nuclear weapon test explosion or any other nuclear explosion."[32] During treaty negotiations, it was difficult for parties at the Conference for Disarmament to agree on what was actually banned. Debates endured about whether or not this language included low-yield testing and subcritical testing. As a consequence, there is much room for interpretation, and parties to the treaty must decide for themselves exactly what counts as a nuclear test. The Clinton administration agreed to adopt a "zero-yield" interpretation of the treaty, meaning that it would agree to ban all nuclear explosions of any yield, but it would reserve the right to conduct subcritical tests.[33] Russia, on the other hand, is rumored to conduct hydro-nuclear tests that do produce a nuclear yield.[34]

If the United States ratifies the CTBT, it will be obligated to adhere to its "zero-yield" interpretation or face severe consequences in the international arena. Adversaries, and even allies who acknowledge a different definition of testing, could continue to conduct nuclear testing and develop new weapons technology, while the United States would be forced to sit idly by and continue its efforts to maintain an outdated stockpile, leaving it with a distinct military disadvantage.

The thought that ratification of the treaty by the United States would have a domino-like effect and induce other nations to follow suit is not as compelling as it sounds. The United States placed a self-imposed moratorium on nuclear explosions in 1992, and it has had virtually no impact on slowing down the advancement of nuclear programs, proliferation, or the testing schedule of some states. Russia and China have continued to make improvements to their nuclear weapons despite Russia's ratification of the treaty and China being a signatory. Russia has even modernized and expanded the types of warheads in its arsenal, and it has revised its doctrine to include the use of nuclear weapons in war to offset its declining conventional forces.[35]

If a desired second order effect of the CTBT is to rally international opposition to proliferation, the moratorium on testing has not had that effect either. Russia has placed new emphasis on its nuclear weapons program, modernizing each leg of its nuclear triad to include new ballistic missile submarines, new heavy intercontinental ballistic missiles (ICBMs), and new low-yield warheads.

Neither has America's ban on testing established a global normative view that all testing is taboo. Since 1992, North Korea has carried out three weapons tests, first in 2006, then 2009, and most recently in 2013. It also appears that North Korea is preparing for its fourth test. Satellite imagery revealed increased activity at its nuclear test site suggesting it is making preparations for another explosion.[36] The fact that preparations are occurring less than a year after its latest test are suggestive of the aggressive nature of its nuclear program. If anything, the United States's moratorium on testing has been a green light to North Korea's nuclear program.

To some extent the US nuclear umbrella has contributed to nonproliferation. Its nuclear arsenal provides a blanket of protection to its non-nuclear allies. US ratification of the CTBT in tandem with new and emerging threats could cause US allies to question the legitimacy and capability of its extended deterrence. Some allies are already questioning the reliability of the US blanket of protection. In response to tensions in the Middle East, Saudi Arabia, and Turkey could make assertive efforts to acquire their own arsenals. One would also expect Japan and South Korea to take up nuclear armament in response to North

Korea's aggressive program. Couple the uncertainty about the health of the aging US stockpile with a permanent ban on testing, and the United States could see its non-nuclear allies taking drastic steps to ensure their national security. Since their conventional forces alone are not enough to guarantee their autonomy, they will naturally turn to nuclear weapons.

As confidence in the US nuclear umbrella erodes, the cooperative relationship the United States shares with its allies may also erode. In effect, ratification of the CTBT could induce proliferation, instead of preventing it. The Strategic Posture Commission report assesses that some US allies believe their security needs can only be met with specific US nuclear capabilities, and the lack of test readiness is viewed as evidence of the decline in the overall commitment of the United States to extended deterrence.[37]

There can never be a guarantee that the actions of the United States to ratify a test ban treaty will prompt other nations to either ratify that same treaty or totally abandon their efforts to adopt a nuclear weapons program. It is hopeful at best to consider that the ratification by the United States would have such a profound diplomatic and symbolic effect. Even if the United States ratified the treaty today it would not enter into force until all of the required 44 states had ratified it. Faced with a prospect where others countries are indeed proliferating, it would not be prudent to assume such a commitment at this time.

Life support for an aging nuclear weapons stockpile

The nuclear weapons in the United States arsenal were designed and built with the specification they would be replaced every 10–15 years. Weapons in the current stockpile are based upon 1970s technology, and the average age of the weapons is 21 years.[38] Built for the Cold War, most of these weapons have long surpassed their intended life cycle and are less relevant to today's emerging threats. They were designed to destroy Soviet hardened targets such as missile silos, and their delivery systems are not consistent with today's expectation of precision-guided systems.[39] Although the LEP and SSP have allowed the weapons to surpass their life expectancy, it has not adapted the arsenal to new and emerging threats.

The moratorium on testing has driven the nuclear enterprise toward a stagnant, maintenance-oriented organization instead of the ambitious and innovative institution it once was. The ban on testing translates into a ban on new weaponry. The majority of nuclear testing was performed for the verification of new weapon designs.[40] The incentive to develop new weapons is obsolete if the weapons cannot be tested. Ratification of the CTBT would end not only US nuclear weapons testing, but also put an end to the development of new nuclear weapons.

Even though the SSP and LEP have extended the life of the weapons in the stockpile, uncertainty about their reliability still looms. Instead of designing and testing new weapons, the arsenal is maintained by an indefinite life support system leaving many to wonder how anyone could have confidence in the stockpile. The weapons systems are a conglomerate of thousands of intricate, precision-crafted parts, and they must interact in an explicit manner in order for the weapon to function.

As the weapons age, plastics become brittle and crack, copper corrodes, and adhesive bonding becomes weak. Environmental factors such as temperature and humidity can impact the rate at which these materials begin to show signs of aging, but being in the presence of uranium and plutonium can greatly accelerate their decaying process.[41] Each of these could lead to catastrophic failure of the weapon, which unfortunately may not be recognized until the weapon is deployed in defense of America's national security. Aging is also seen in the plutonium core, the most important part of the warhead, which is responsible for the warhead's explosive power. Albeit a very slow process, plutonium will decay over time losing both mass and energy, and the bomb will either detonate with a lower yield than intended or not detonate at all.[42]

Likewise, the weapon systems are so precise, that one cannot be certain how a 20-plus-year-old weapon will perform. Without testing, no one can say with certainty that either the SSP or the LEP program is a success. The most definitive demonstration of reliability is an explosive test. This would be comparable to placing a patient on a life support system that has not been turned on in 20 years.

Modernization of the stockpile would have several advantages. First, it would allow the US to incorporate relevant safety and security features into the weapon systems and provide the capability to tailor the weapons to support the contemporary strategic environment. Second, the longer the delay in modernization, the higher the price tag will be. A recent study indicates the cost could be upwards of $352 billion over the next decade.[43] Third, modernization efforts would help maintain military effectiveness and reinforce the nation's commitment to its extended nuclear deterrent obligations. This brings along the added benefit of improving international relationships and stalling nuclear acquisitions of non-nuclear allies. Finally, a modernization program would enable the National Laboratories to recruit and retain technical expertise. The absence of testing and the lack of new weapons designs have left the enterprise with a deficit of experienced personnel. The SSP has directly contributed to a decline in science and engineering capabilities. Instead of being able to follow a weapon through its entire life cycle, engineers and scientists have become adept at analyzing and defining small variances in the stockpile based upon knowledge obtained, primarily, four or five decades ago when the United States was actively testing and developing new weapons.[44] This presents a real challenge if the CTBT is ratified as the incentive to retain such expertise would be lost.

Although scientists would argue that computerized modeling has provided them with a better understanding of the weapons systems, the simulated tests can never fully replicate the knowledge gained from live, explosive testing. The information used in the virtual testing ground is comprised of information gained from previous explosive tests and guesswork. In designing the computer codes, scientists had to hypothesize and theorize about missing data points, thus leaving a margin of uncertainty in the simulated results.[45] Furthermore, it is risky to rely on the simulated tests as an indicator of reliability, because the aging weapons have undergone significant changes due to degradation and replacement of key components under LEP. Testing is the best way to ensure that repairs to the weapons have resolved known problems.

The verification regime: a pipedream

Article IV of the CTBT establishes the verification regime, which includes on-site inspections and the IMS. Even with advances in the IMS, the treaty lacks any enforceable mechanisms. If an explosion is detected, a request can be submitted for an on-site inspection. However, a decision on whether or not to approve the request is required within 96 hours, and it must have a consensus of at least 30 affirmative votes from members of the executive council. Gathering a consensus on that level could be a diplomatic and political nightmare. The fact that "testing" is not explicitly defined by the treaty would challenge efforts for granting an on-site inspection, because the notion of what constitutes a nuclear test could be disputed. Thus, on-site inspections are not easily achievable, and therefore there is no guarantee that a nuclear explosion could be verified in the manner established by the CTBT.

If an inspection were granted, several more challenges would have to be overcome. Determining the exact location of the suspected explosion would be difficult. If the gases have vented or the explosion was not conducted to a certain underground depth, monitoring equipment would not be able to pinpoint the exact location.[46] Additionally, the treaty mandates that the inspection report must be transmitted to the CTBTO Executive Council no later than 25 days after the approval of the inspection. It is unrealistic to expect that a quorum of at least 30 member states would agree to the inspection, dispatch an inspection team, find the detonation site, gather and analyze the data, and submit a report in less than a month.

An additional challenge would be seen by those who would try to explode a device but escape detection. The seismic signal of an explosion can be reduced to a level below detection by launching the explosion in a deep cavity located in high-strength, low-porosity rock or in salt.[47] Careful selection of a testing site can offer a high level of confidence that the test will go undetected. For instance, the IMS was aware of North Korea's impending 2009 test, and even though North Korea made no attempt to hide it, the IMS did not detect any radionuclides following the explosion.[48] Another approach to cheating would be to simply explode a bomb without attribution. This scenario entails the placement of a nuclear device either on or in the ocean. The device can subsequently be detonated hours or days later. It would theoretically be detected by the IMS, but the guilty party could simply deny its culpability.[49]

Other challenges to the IMS come in the form of environmental and administrative obstacles. Site selection for an IMS station is very discriminatory. For example, the location for a seismic station must be evaluated for vibrations caused by "wind, surf, traffic, and so on to ensure it is quiet enough for the station to be a good detector of seismic events."[50] For this reason, IMS stations are often located in remote, difficult to traverse locations. Also, once the station is operational it will require maintenance, upgrades, or replacement as new technologies emerge. It can be a burdensome task to get to the station for routine maintenance. Additionally, administrative procedures within the host country often have to be negotiated through several agencies within each country before the site can be approved. Once the host country is ready to proceed with the station, the CTBT requires a legally binding Facility Agreement between the host country and the CTBTO that grants the latter legal and administrative authority to conduct its work at the station.[51] Aside from the obstacles encountered in setting up a monitoring station, the IMS and verification regime cannot be fully operational until the treaty is entered into force. This would mean that countries such as India and Pakistan would have to ratify the treaty before IMS stations could be set up within their boundaries. It also means that the CTBTO has to coordinate its efforts and monitoring network in 89 different countries. Given all the caveats associated with activating the IMS, it is unlikely it will ever be fully operational.

Conclusion and recommendations

While it is important for the United States to engage with the international community on the subject of nuclear arms control, US policymakers would serve the country well to remember that deterrence is very much dependent upon credibility. There is no better way to demonstrate the proficiency of the US nuclear enterprise and reinforce confidence than through an explosive test. Therefore, the first viable option is to, at a minimum, resume a rigorous underground nuclear testing schedule. Scientists and politicians can debate forever as to how to adequately define nuclear surety. However, nuclear weapons have been central to the ability of the US to maintain its international commitments and to avoid the risk of nuclear coercion against its allies.

In the past, nuclear experiments were used to assess and evaluate the behavior of the nuclear warhead and the properties of different materials used in the weapon. These experiments were invaluable for obtaining data that could be used in the development of new weapons or to assist in the design of future tests. Therefore, a second option would be to implement a limited testing schedule to provide scientists with the opportunity to gather data. For example, an underground test every three to five years would net a wealth of data. It would help to ensure that scientists understand the behavior of different weapons designs, instill confidence in the computer models used to predict the behavior of the weapons, and ultimately assess the effects of age-related changes in the current stockpile.

Although ratification of the CTBT might assist the US in meeting its nonproliferation objectives, the risk of depreciating US defense credibility is too great, especially with an aging stockpile of questionable integrity. Treaties require a high degree of transparency and cooperation. The CTBT, a multilateral and multinational treaty, elevates not only the level of cooperation required, but it may also jeopardize the individual interests among the diverse nations.

For decades nuclear weapons have been an integral part of the national defense of the United States. Despite the best efforts of the SSP and LEP, the arsenal continues to age, and at some point in the future it will reach a time when safety and reliability can no longer be assured. By that time the nuclear enterprise will be seriously deficient in expertise and infrastructure. Ratification of the CTBT would undermine not only the national security of the United States, but also that of its allies.

Appendix

Table A11.1 Status of signature and ratification

States	Signature	Ratification
Afghanistan	24-SEP-2003	24-SEP-2003
Albania	27-SEP-1996	23-APR-2003
Algeria*	15-OCT-1996	11-JUL-2003
Andorra	24-SEP-1996	12-JUL-2006
Angola	27-SEP-1996	
Antigua and Barbuda	16-APR-1997	11-JAN-2006
Argentina*	24-SEP-1996	04-DEC-1998
Armenia	01-OCT-1996	12-JUL-2006
Australia*	24-SEP-1996	09-JUL-1998
Austria*	24-SEP-1996	13-MAR-1998
Azerbaijan	28-JUL-1997	02-FEB-1999
Bahamas	04-FEB-2005	30-NOV-2007
Bahrain	24-SEP-1996	12-APR-2004
Bangladesh*	24-OCT-1996	08-MAR-2000
Barbados	14-JAN-2008	14-JAN-2008
Belarus	24-SEP-1996	13-SEP-2000
Belgium*	24-SEP-1996	29-JUN-1999
Belize	14-NOV-2001	26-MAR-2004
Benin	27-SEP-1996	06-MAR-2001
Bhutan		
Bolivia	24-SEP-1996	04-OCT-1999
Bosnia and Herzegovina	24-SEP-1996	26-OCT-2006
Botswana	16-SEP-2002	28-OCT-2002
Brazil*	24-SEP-1996	24-JUL-1998
Brunei Darussalam	22-JAN-1997	10-JAN-2013
Bulgaria*	24-SEP-1996	29-SEP-1999
Burkina Faso	27-SEP-1996	17-APR-2002
Burundi	24-SEP-1996	24-SEP-2008
Cambodia	26-SEP-1996	10-NOV-2000
Cameroon	16-NOV-2001	06-FEB-2006
Canada*	24-SEP-1996	18-DEC-1998
Cape Verde	01-OCT-1996	01-MAR-2006
Côte d'Ivoire	25-SEP-1996	11-MAR-2003

continued

States	Signature	Ratification
Central African Republic	19-DEC-2001	26-MAY-2010
Chad	08-OCT-1996	08-FEB-2013
Chile*	24-SEP-1996	12-JUL-2000
China*	24-SEP-1996	
Colombia*	24-SEP-1996	29-JAN-2008
Comoros	12-DEC-1996	
Congo	11-FEB-1997	
Cook Islands	05-DEC-1997	06-SEP-2005
Costa Rica	24-SEP-1996	25-SEP-2001
Croatia	24-SEP-1996	02-MAR-2001
Cuba		
Cyprus	24-SEP-1996	18-JUL-2003
Czech Republic	12-NOV-1996	11-SEP-1997
Democratic People's Republic of Korea*		
Democratic Republic of the Congo*	04-OCT-1996	28-SEP-2004
Denmark	24-SEP-1996	21-DEC-1998
Djibouti	21-OCT-1996	15-JUL-2005
Dominica		
Dominican Republic	03-OCT-1996	04-SEP-2007
Ecuador	24-SEP-1996	12-NOV-2001
Egypt*	14-OCT-1996	
El Salvador	24-SEP-1996	11-SEP-1998
Equatorial Guinea	09-OCT-1996	
Eritrea	11-NOV-2003	11-NOV-2003
Estonia	20-NOV-1996	13-AUG-1999
Ethiopia	25-SEP-1996	08-AUG-2006
Fiji	24-SEP-1996	10-OCT-1996
Finland*	24-SEP-1996	15-JAN-1999
France*	24-SEP-1996	06-APR-1998
Gabon	07-OCT-1996	20-SEP-2000
Gambia	09-APR-2003	
Georgia	24-SEP-1996	27-SEP-2002
Germany*	24-SEP-1996	20-AUG-1998
Ghana	03-OCT-1996	14-JUN-2011
Greece	24-SEP-1996	21-APR-1999
Grenada	10-OCT-1996	19-AUG-1998
Guatemala	20-SEP-1999	12-JAN-2012
Guinea	03-OCT-1996	20-SEP-2011
Guinea-Bissau	11-APR-1997	24-SEP-2013
Guyana	07-SEP-2000	07-MAR-2001
Haiti	24-SEP-1996	01-DEC-2005
Holy See	24-SEP-1996	18-JUL-2001
Honduras	25-SEP-1996	30-OCT-2003
Hungary*	25-SEP-1996	13-JUL-1999
Iceland	24-SEP-1996	26-JUN-2000
India*		
Indonesia*	24-SEP-1996	06-FEB-2012
Iran (Islamic Republic of)*	24-SEP-1996	
Iraq	19-AUG-2008	26-SEP-2013
Ireland	24-SEP-1996	15-JUL-1999

States	Signature	Ratification
Israel*	25-SEP-1996	
Italy*	24-SEP-1996	01-FEB-1999
Jamaica	11-NOV-1996	13-NOV-2001
Japan*	24-SEP-1996	08-JUL-1997
Jordan	26-SEP-1996	25-AUG-1998
Kazakhstan	30-SEP-1996	14-MAY-2002
Kenya	14-NOV-1996	30-NOV-2000
Kiribati	07-SEP-2000	07-SEP-2000
Kuwait	24-SEP-1996	06-MAY-2003
Kyrgyzstan	08-OCT-1996	02-OCT-2003
Lao People's Democratic Republic	30-JUL-1997	05-OCT-2000
Latvia	24-SEP-1996	20-NOV-2001
Lebanon	16-SEP-2005	21-NOV-2008
Lesotho	30-SEP-1996	14-SEP-1999
Liberia	01-OCT-1996	17-AUG-2009
Libya	13-NOV-2001	06-JAN-2004
Liechtenstein	27-SEP-1996	21-SEP-2004
Lithuania	07-OCT-1996	07-FEB-2000
Luxembourg	24-SEP-1996	26-MAY-1999
Madagascar	09-OCT-1996	15-SEP-2005
Malawi	09-OCT-1996	21-NOV-2008
Malaysia	23-JUL-1998	17-JAN-2008
Maldives	01-OCT-1997	07-SEP-2000
Mali	18-FEB-1997	04-AUG-1999
Malta	24-SEP-1996	23-JUL-2001
Marshall Islands	24-SEP-1996	28-OCT-2009
Mauritania	24-SEP-1996	30-APR-2003
Mauritius		
Mexico*	24-SEP-1996	05-OCT-1999
Micronesia, Federated States of	24-SEP-1996	25-JUL-1997
Monaco	01-OCT-1996	18-DEC-1998
Mongolia	01-OCT-1996	08-AUG-1997
Montenegro	23-OCT-2006	23-OCT-2006
Morocco	24-SEP-1996	17-APR-2000
Mozambique	26-SEP-1996	04-NOV-2008
Myanmar, Republic of the Union of	25-NOV-1996	
Namibia	24-SEP-1996	29-JUN-2001
Nauru	08-SEP-2000	12-NOV-2001
Nepal	08-OCT-1996	
Netherlands*	24-SEP-1996	23-MAR-1999
New Zealand	27-SEP-1996	19-MAR-1999
Nicaragua	24-SEP-1996	05-DEC-2000
Niger	03-OCT-1996	09-SEP-2002
Nigeria	08-SEP-2000	27-SEP-2001
Niue	09-APR-2012	
Norway*	24-SEP-1996	15-JUL-1999
Oman	23-SEP-1999	13-JUN-2003
Pakistan*		

continued

States	Signature	Ratification
Palau	12-AUG-2003	01-AUG-2007
Panama	24-SEP-1996	23-MAR-1999
Papua New Guinea	25-SEP-1996	
Paraguay	25-SEP-1996	04-OCT-2001
Peru*	25-SEP-1996	12-NOV-1997
Philippines	24-SEP-1996	23-FEB-2001
Poland*	24-SEP-1996	25-MAY-1999
Portugal	24-SEP-1996	26-JUN-2000
Qatar	24-SEP-1996	03-MAR-1997
Republic of Korea*	24-SEP-1996	24-SEP-1999
Republic of Moldova	24-SEP-1997	16-JAN-2007
Romania*	24-SEP-1996	05-OCT-1999
Russian Federation*	24-SEP-1996	30-JUN-2000
Rwanda	30-NOV-2004	30-NOV-2004
Saint Kitts and Nevis	23-MAR-2004	27-APR-2005
Saint Lucia	04-OCT-1996	05-APR-2001
Saint Vincent and the Grenadines	02-JUL-2009	23-SEP-2009
Samoa	09-OCT-1996	27-SEP-2002
San Marino	07-OCT-1996	12-MAR-2002
Sao Tome and Principe	26-SEP-1996	
Saudi Arabia		
Senegal	26-SEP-1996	09-JUN-1999
Serbia	08-JUN-2001	19-MAY-2004
Seychelles	24-SEP-1996	13-APR-2004
Sierra Leone	08-SEP-2000	17-SEP-2001
Singapore	14-JAN-1999	10-NOV-2001
Slovakia*	30-SEP-1996	03-MAR-1998
Slovenia	24-SEP-1996	31-AUG-1999
Solomon Islands	03-OCT-1996	
Somalia		
South Africa*	24-SEP-1996	30-MAR-1999
South Sudan		
Spain*	24-SEP-1996	31-JUL-1998
Sri Lanka	24-OCT-1996	
Sudan	10-JUN-2004	10-JUN-2004
Suriname	14-JAN-1997	07-FEB-2006
Swaziland	24-SEP-1996	
Sweden*	24-SEP-1996	02-DEC-1998
Switzerland*	24-SEP-1996	01-OCT-1999
Syrian Arab Republic		
Tajikistan	07-OCT-1996	10-JUN-1998
Thailand	12-NOV-1996	
The former Yugoslav Republic of Macedonia	29-OCT-1998	14-MAR-2000
Timor-Leste	26-SEP-2008	
Togo	02-OCT-1996	02-JUL-2004
Tonga		
Trinidad & Tobago	08-OCT-2009	26-MAY-2010
Tunisia	16-OCT-1996	23-SEP-2004
Turkey*	24-SEP-1996	16-FEB-2000
Turkmenistan	24-SEP-1996	20-FEB-1998

States	Signature	Ratification
Tuvalu		
Uganda	07-NOV-1996	14-MAR-2001
Ukraine*	27-SEP-1996	23-FEB-2001
United Arab Emirates	25-SEP-1996	18-SEP-2000
United Kingdom of Great Britain and Northern Ireland*	24-SEP-1996	06-APR-1998
United Republic of Tanzania	30-SEP-2004	30-SEP-2004
United States of America*	24-SEP-1996	
Uruguay	24-SEP-1996	21-SEP-2001
Uzbekistan	03-OCT-1996	29-MAY-1997
Vanuatu	24-SEP-1996	16-SEP-2005
Venezuela (Bolivarian Republic of)	03-OCT-1996	13-MAY-2002
Vietnam*	24-SEP-1996	10-MAR-2006
Yemen	30-SEP-1996	
Zambia	03-DEC-1996	23-FEB-2006
Zimbabwe	13-OCT-1999	

Note
* Signature and ratification required for treaty to enter into force.

The CTBT will enter into force after the 44 designated states have ratified the treaty. As of September 26, 2013,[53] 182 states have signed the treaty and 151 have ratified the treaty. Of the 44 specified countries, only 35 have ratified the treaty.

Notes

1 Trinity Atomic website, www.cddc.vt.edu/host/atomic/trinity/trinity1.html.
2 Trinity Atomic website.
3 David A. Koplow, *Testing a Nuclear Test Ban: What Should Be Prohibited by a Comprehensive Treaty?* (Brookfield, VT: Dartmouth Publishing Company), 8.
4 Jonathan Medalia, "Comprehensive Nuclear Test Ban Treaty: Background and Current Developments," Congressional Research Service, November 15, 2006, 2.
5 Global Solutions, "About the Comprehensive Test-Ban Treaty," http://globalsolutions.org/prevent-war/nuclear-disarmament/comprehensive-test-ban-treaty-ctbt.
6 Medalia, "Comprehensive Nuclear Test Ban Treaty," 2.
7 Comprehensive Nuclear Test Ban Treaty Organization, "The Treaty: The Comprehensive Nuclear Test Ban Treaty," www.ctbto.org/the-treaty/.
8 Comprehensive Nuclear Test Ban Treaty Organization, "The Treaty: The Comprehensive Nuclear Test Ban Treaty."
9 Tom Z. Collina and Daryl G. Kimball, "Now More Than Ever: The Case for the Comprehensive Nuclear Test Ban Treaty," Arms Control Association, February 2010, 1, www.armscontrol.org/system/files/ACA_CTB_Briefing_Book.pdf.
10 Lt. Col. Jeffrey D. Neischel, "The Comprehensive Nuclear Test Ban Treaty," USAF Counterproliferation Center, Counterproliferation Paper No. 54, June 2010, 24.
11 Neischel, "The Comprehensive Nuclear Test Ban Treaty," 24.
12 George Bunn and Christopher F. Chyba, *US Nuclear Weapons Policy: Confronting Today's Threats* (Baltimore, MD: Brookings Institution Press, 2006), 249.
13 Medalia, "Comprehensive Nuclear Test Ban Treaty," 3.

14 National Nuclear Security Administration, "Managing the Stockpile," http://nnsa. energy.gov/ourmission/managingthestockpile/lifeextensionprogram.
15 National Nuclear Security Administration, "Managing the Stockpile."
16 National Nuclear Security Administration, "Managing the Stockpile."
17 US Department of State, "Key Accomplishments of the Stockpile Stewardship Program," Bureau of Arms Control Fact Sheet, April 11, 2012, www.state.gov/t/avc/rls/187694.htm.
18 Jonathan Medalia, "Nuclear Warheads: The Reliable Replacement Warhead Program and the Life Extension Program," Congressional Research Service, December 3, 2007, 6.
19 US Department of State, "Key Accomplishments of the Stockpile Stewardship Program."
20 Arnie Heller, "Annual Certification Takes a Snapshot of Stockpile's Health: The Nuclear Arsenal Gets a Yearly Checkup," *Science and Technology Review* (July/August 2001): 5.
21 Heller, "Annual Certification Takes a Snapshot of Stockpile's Health," 6.
22 Rebecca Boyle, "Advanced Supercomputer Models Supplant Real-World Nuclear Weapons Tests, But Are They Accurate?" *Popular Science*, November 2, 2011, www.popsci.com/technology/article/2011-11/advanced-supercomputer-models-supplant-nuclear-weapons-tests-are-they-accurate.
23 Clay Dillingham, "But Will It Work?" *National Security Science* (April 2013): 4, www.lanl.gov/science/NSS/pdf/NSS_April_2013.pdf.
24 Ann Parker, "New Day Dawns in Supercomputing," *Science and Technology Review* (June 2000), www.llnl.gov/str/Seager.html.
25 Donald B. Johnston, "Sequoia Supercomputer Transitions to Classified Work," Lawrence Livermore National Laboratory, April 2013, https://www.llnl.gov/news/aroundthelab/2013/Apr/ATL041713_sequoia.html.
26 Eileen Patterson, "Roadrunner on the Way to Trinity," *National Security Science* (April 2013), www.lanl.gov/science/NSS/pdf/NSS_April_2013.pdf.
27 United Nations, "Ending Nuclear Testing," www.un.org/en/events/againstnucleartestsday/history.shtml#a31.
28 Comprehensive Nuclear Test Ban Treaty Organization, "The Global Verification Regime and the International Monitoring System," www.nuclearfiles.org/menu/key-issues/nuclear-weapons/issues/arms-control-disarmament/verification/booklet3.pdf.
29 Medalia, "Comprehensive Nuclear Test Ban Treaty," 19.
30 Comprehensive Nuclear Test Ban Treaty Organization, "Nuclear Testing," www.ctbto.org/nuclear-testing/history-of-nuclear-testing/.
31 Amrutha Gayathri, "Pakistan Could Deliver Nuclear Weapons to Saudi Arabia If Iran Acquires Nukes," *International Business Times*, November 7, 2013, www.ibtimes.com/pakistan-could-deliver-nuclear-weapons-saudi-arabia-if-iran-acquires-nukes-report-1459298.
32 Keith A. Hansen, *The Comprehensive Nuclear Test Ban Treaty: An Insider's Perspective* (Stanford, CA: Stanford University Press, 2006), 88.
33 Hansen, *The Comprehensive Nuclear Test Ban Treaty: An Insider's Perspective*, 25.
34 Hansen, *The Comprehensive Nuclear Test Ban Treaty: An Insider's Perspective*, 25.
35 World Affairs Council, "The Nuclear Forces and Doctrine of the Russian Federation and the People's Republic of China," October 2011, 5, www.worldaffairscouncils.org/2011/images/insert/Majority%20Statement%20and%20Testimony.pdf.
36 World Affairs Council, "The Nuclear Forces and Doctrine of the Russian Federation and the People's Republic of China," 5.
37 Eun-jung Kim, "North Korea Ready for 4th Nuclear Test, Yet No Imminent Sign," *Global Post*, November 15, 2013, www.globalpost.com/dispatch/news/yonhap-news-agency/131115/n-korea-ready-4th-nuclear-test-yet-no-imminent-sign-seoul.

38 Adm. Richard W. Mies, "Strategic Deterrence in the 21st Century," *National Security Science* (April 2013), www.lanl.gov/newsroom/publications/national-security-science/2013-april/_assets/docs/strategic-deterrence.pdf.

39 Mies, "Strategic Deterrence in the 21st Century."

40 Allison Bond, "Battling over Aging Nuclear Warheads," *Popular Science*, April 15, 2009, www.popsci.com/military-aviation-amp-space/2009–04/battling-over-aging-nuclear-warheads.

41 Collina and Kimball, "Now More Than Ever," 8.

42 Dillingham, "But Will It Work?" 3.

43 Collina and Kimball, "Now More Than Ever," 11.

44 Dana Priest, "Aging U.S. Nuclear Arsenal Slated for Costly and Long-Delayed Modernization," *Washington Post*, September 15, 2012, http://articles.washingtonpost.com/2012-09-15/world/35497119_1_nuclear-stockpile-nuclear-weapons-nuclear-facilities.

45 Kathleen C. Bailey, "The Comprehensive Test Ban Treaty: An Update on the Debate," National Institute for Public Policy, 2001, 14, www.nipp.org/National%20Institute%20Press/Archives/Publication%20Archive%20PDF/CTBT%20Update.pdf.

46 Sidney D. Drell and Marvin L. Adams, "Technical Issues in Keeping the Nuclear Stockpile Safe, Secure, and Reliable," American Association for the Advancement of Science, www.aaas.org/cstsp/files/DrellAdamsBrief.pdf.

47 Bailey, "Comprehensive Nuclear Test Ban Treaty," 23.

48 Bailey, "Comprehensive Nuclear Test Ban Treaty," 23.

49 Comprehensive Nuclear Test Ban Treaty Organization, "Experts Sure about Nature of the DPRK Event," www.ctbto.org/press-centre/highlights/2009/experts-sure-about-nature-of-the-dprk-event/.

50 Bailey, "Comprehensive Nuclear Test Ban Treaty," 24.

51 Comprehensive Nuclear Test Ban Treaty Organization, "The Global Verification Regime."

Bibliography

Bailey, Kathleen C., "The Comprehensive Test Ban Treaty: An Update on the Debate," National Institute for Public Policy, 2001, www.nipp.org/National%20Institute%20Press/Archives/Publication%20Archive%20PDF/CTBT%20Update.pdf.

Bond, Allison, "Battling over Aging Nuclear Warheads," *Popular Science*, April 15, 2009, www.popsci.com/military-aviation-amp-space/2009-04/battling-over-aging-nuclear-warheads.

Boyle, Rebecca, "Advanced Supercomputer Models Supplant Real-World Nuclear Weapons Tests, But Are They Accurate?" *Popular Science*, November 2, 2011, www.popsci.com/technology/article/2011-11/advanced-supercomputer-models-supplant-nuclear-weapons-tests-are-they-accurate.

Bunn, George, and Chyba, Christopher F., *US Nuclear Weapons Policy: Confronting Today's Threats* (Baltimore, MD: Brookings Institution Press, 2006).

Collina, Tom Z., and Kimball, Daryl G., "Now More Than Ever: The Case for the Comprehensive Nuclear Test Ban Treaty," Arms Control Association, February 2010, www.armscontrol.org/system/files/ACA_CTB_Briefing_Book.pdf.

Comprehensive Nuclear Test Ban Treaty Organization, "Experts Sure About Nature of the DPRK Event," www.ctbto.org/press-centre/highlights/2009/experts-sure-about-nature-of-the-dprk-event/.

Comprehensive Nuclear Test Ban Treaty Organization, "Nuclear Testing," www.ctbto.org/nuclear-testing/history-of-nuclear-testing/.

Comprehensive Nuclear Test Ban Treaty Organization, "The Global Verification Regime and the International Monitoring System," www.nuclearfiles.org/menu/key-issues/nuclear-weapons/issues/arms-control-disarmament/verification/booklet3.pdf.

Comprehensive Nuclear Test Ban Treaty Organization, "The Treaty: The Comprehensive Nuclear Test Ban Treaty," www.ctbto.org/the-treaty/.

Dillingham, Clay, "But Will It Work?" *National Security Science* (April 2013), www.lanl.gov/science/NSS/pdf/NSS_April_2013.pdf.

Drell, Sidney D., and Adams, Marvin L., "Technical Issues in Keeping the Nuclear Stockpile Safe, Secure, and Reliable," American Association for the Advancement of Science, www.aaas.org/cstsp/files/DrellAdamsBrief.pdf.

Gayathri, Amrutha, "Pakistan Could Deliver Nuclear Weapons to Saudi Arabia If Iran Acquires Nukes," *International Business Times*, November 7, 2013, www.ibtimes.com/pakistan-could-deliver-nuclear-weapons-saudi-arabia-if-iran-acquires-nukes-report-1459298.

Global Solutions, "About the Comprehensive Test-Ban Treaty," http://globalsolutions.org/prevent-war/nuclear-disarmament/comprehensive-test-ban-treaty-ctbt.

Hansen, Keith A., *The Comprehensive Nuclear Test Ban Treaty: An Insider's Perspective* (Stanford, CA: Stanford University Press, 2006).

Heller, Arnie, "Annual Certification Takes a Snapshot of Stockpile's Health: The Nuclear Arsenal Gets a Yearly Checkup," *Science and Technology Review* (July/August 2001).

Koplow, David A., *Testing a Nuclear Test Ban: What Should Be Prohibited by a Comprehensive Treaty?* (Brookfield, VT: Dartmouth Publishing Company).

Johnston, Donald B., "Sequoia Supercomputer Transitions to Classified Work," Lawrence Livermore National Laboratory, April 2013, www.llnl.gov/news/aroundthelab/2013/Apr/ATL041713_sequoia.html.

Kim, Eun-jung, "North Korea Ready for 4th Nuclear Test, Yet No Imminent Sign," *Global Post*, November 15, 2013, www.globalpost.com/dispatch/news/yonhap-news-agency/131115/n-korea-ready-4th-nuclear-test-yet-no-imminent-sign-seoul.

Medalia, Jonathan, "Comprehensive Nuclear Test Ban Treaty: Background and Current Developments," Congressional Research Service, November 15, 2006.

Medalia, Jonathan, "Nuclear Warheads: The Reliable Replacement Warhead Program and the Life Extension Program," Congressional Research Service, December 3, 2007.

Mies, Adm. Richard W., "Strategic Deterrence in the 21st Century," *National Security Science* (April 2013), www.lanl.gov/newsroom/publications/national-security-science/2013-april/_assets/docs/strategic-deterrence.pdf.

National Nuclear Security Administration, "Managing the Stockpile," http://nnsa.energy.gov/ourmission/managingthestockpile/lifeextensionprogram.

Neischel, Lt. Col. Jeffrey D., "The Comprehensive Nuclear Test Ban Treaty," USAF Counterproliferation Center, Counterproliferation Paper No. 54, June 2010.

Parker, Ann, "New Day Dawns in Supercomputing," *Science and Technology Review* (June 2000), www.llnl.gov/str/Seager.html.

Patterson, Eileen, "Roadrunner on the Way to Trinity," *National Security Science* (April 2013), www.lanl.gov/science/NSS/pdf/NSS_April_2013.pdf.

Priest, Dana, "Aging U.S. Nuclear Arsenal Slated for Costly and Long-Delayed Modernization," *Washington Post*, September 15, 2012, http://articles.washingtonpost.com/2012-09-15/world/35497119_1_nuclear-stockpile-nuclear-weapons-nuclear-facilities.

Trinity Atomic website, www.cddc.vt.edu/host/atomic/trinity/trinity1.html.

United Nations, "Ending Nuclear Testing," www.un.org/en/events/againstnucleartests-day/history.shtml#a31.

US Department of State, "Key Accomplishments of the Stockpile Stewardship Program," Bureau of Arms Control Fact Sheet, April 11, 2012, www.state.gov/t/avc/rls/187694. htm.

World Affairs Council, "The Nuclear Forces and Doctrine of the Russian Federation and the People's Republic of China," October 2011, www.worldaffairscouncils.org/2011/ images/insert/Majority%20Statement%20and%20Testimony.pdf.

12 Nuclear arms reductions after New START

Obstacles and options

Stephen J. Cimbala

Introduction

As US President Barack Obama approached the midpoint of his second term in office, the political atmosphere for Russo-American nuclear arms reductions was fraught and seemingly unfavorable for meaningful progress. Russia's March 2014 annexation of Crimea and subsequent destabilization of eastern Ukraine capped the Obama "reset" in US–Russian relations and created a palpable feeling of distrust between NATO and Russia.[1] In addition, Russian President Vladimir Putin explicitly introduced the threat of nuclear escalation into the crisis in Ukraine. As Russian troops and military equipment continued to flow into eastern Ukraine in support of pro-Russian separatists there, Putin issued a public warning in late August 2014: "I want to remind you that Russia is one of the most powerful nuclear nations," and added that "this is a reality, not just words."[2] Putin also indicated that Russia was strengthening its nuclear deterrent forces and noted that "it's best not to mess with us" over Ukraine.[3] In mid-September 2014 Putin warned that Russia would counter military moves by the United States and NATO with nuclear and conventional force modernization.[4] Earlier in the same month, a Russian general and senior Defense Ministry official called for Russia to revise its military doctrine and specify conditions under which Russia might launch a preemptive nuclear strike against NATO.[5]

On the other hand, the United States and Russia cannot escape certain of their shared responsibilities without endangering their own national interests. One of these shared responsibilities is for the management of nuclear world order. The bipolar nuclear world of the Cold War has given way to an emerging international system of nuclear indeterminacy and uncertainty. Without US and Russian leadership in nuclear arms control, devolution of the current nonproliferation regime into a hodgepodge of regional nuclear and nuclear-aspiring actors, with disparate nationalisms stoking local political rivalries, is not inconceivable. Instead of nuclear deterrence and arms control stability, various regions may be marked by nuclear "fracking" as additional actors build their own nuclear weapons capabilities or have them provided by outsiders. At a certain unknown tipping point, the system-supporting props of a viable nonproliferation regime and of US–Russian leadership in nuclear arms control might give way to "do it yourself" franchises in nuclear entrepreneurship, and not only among states. The discussion that follows examines options for post-New START (Strategic Arms Reduction Treaty) strategic nuclear arms reductions by the United States and Russia, notwithstanding disputes over Ukraine and other contentious issues.

After New START

The New START agreement of 2010 (entering into effect in 2011) requires that Russia and the United States reduce their numbers of operationally deployed warheads on intercontinental launchers (land- and sea-based strategic missiles and heavy bombers) to no more than 1,550 deployed warheads and 700 deployed launchers by 2018.[6] Each state deploys a triad of long-range nuclear-launchers, and both the United States and Russia have plans for nuclear force modernization. In the analysis that follows, we generated hypothetical, but not unrealistic, US and Russian New START-compliant strategic nuclear forces for the period 2018–2020. These forces were then subjected to first strikes and their numbers of surviving and retaliating second-strike warheads were estimated under each of four conditions of operational readiness and launch preparedness: (1) generated alert, and forces are launched on warning; (2) generated alert, and forces are launched after riding out the attack; (3) day-to-day alert, and forces are launched on warning; and (4) day-to-day alert, and forces ride out the attack.[7] The results of this analysis for the two states' surviving and retaliating forces under a peacetime deployment limit of 1,550 warheads for each state appear in Figure 12.1, below.

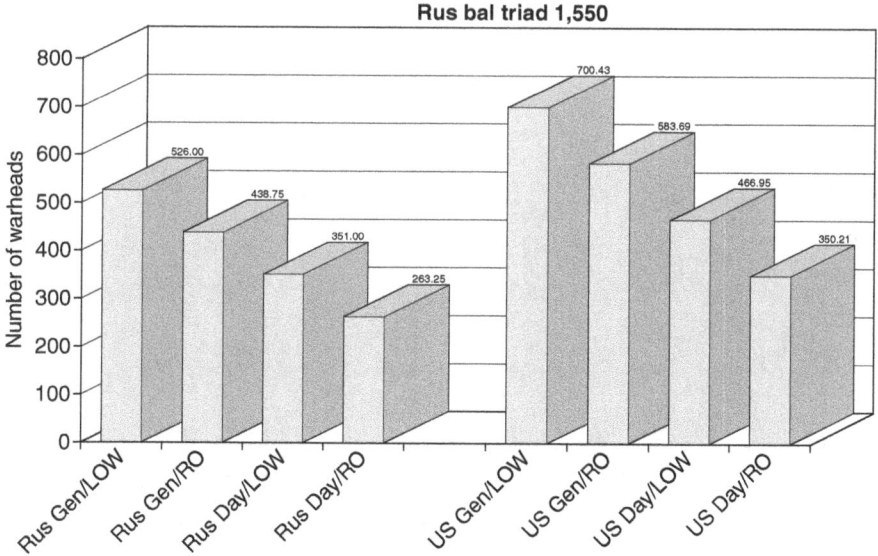

Figure 12.1 US–Russia surviving and retaliating warheads 1,550 deployment limit

The data summarized in Figure 12.1 show that US and Russian forces with peacetime deployment limits of 1,550 weapons on intercontinental launchers can meet the requirements, not only for assured retaliation, but also for flexible use against a variety of target sets including opposing forces, command-control facilities, and urban-industrial targets. In the canonical case of "generated alert, riding out the attack" each state can also withhold some retaliating weapon for future strikes as a means of intra-war deterrence and support for war termination. These numbers of surviving weapons can also support adaptability, in the sense of flexibility and resilience, for surviving weapons and launchers. Can the same be said about post-New START forces with smaller numbers of peacetime deployed weapons?

An incremental approach

Two possibilities exist for future post-New START reductions in US and Russian strategic nuclear forces. One approach would be incremental. In this approach, each state would reduce its number of deployed long-range nuclear weapons to a maximum of 1,000 or so. Under the assumptions that each state deploys a maximum number of 1,000 weapons, the outcomes for each in terms of second-strike surviving and retaliating weapons are summarized in Figure 12.2, as below.

As might be expected, the numbers of surviving and retaliating weapons for the US and for Russia are smaller than they were in the previous case of 1,550 deployed weapons. But that finding does not tell the entire story. It is also the

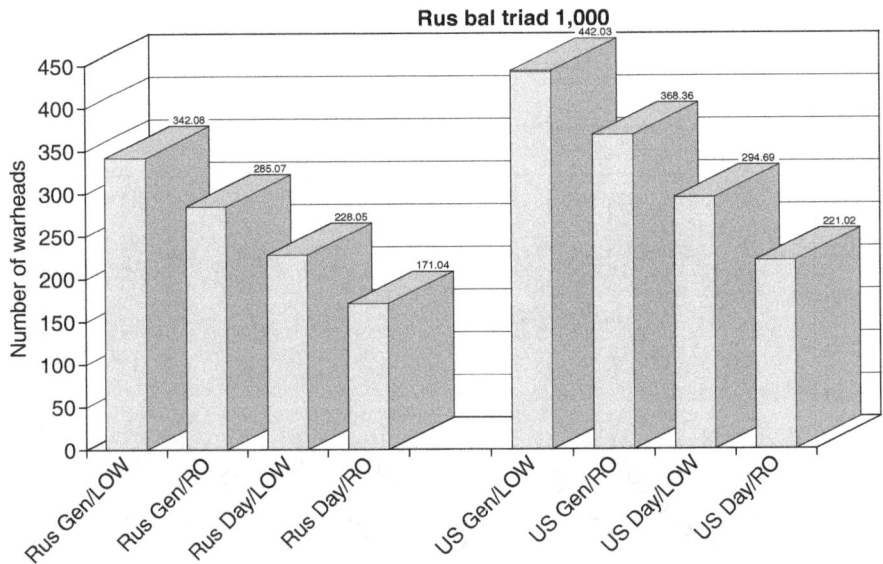

Figure 12.2 US–Russia surviving and retaliating warheads 1,000 deployment limit

case that some degrees of freedom in other areas are lost. The numbers of second-strike surviving and retaliating warheads for each state in the case of 1,000 prewar deployed weapons, compared to the 1,550 case, are more restrictive of flexible targeting options, of operational withholds for follow-on attacks, and for the retention of residual forces to support post-attack escalation control or war termination. Under some conditions, especially on day-to-day alert, either state might be challenged to fulfill the requirements of the assured retaliation mission promptly if unexpected technical glitches encumbered either the launch or command-control systems.

The case of a deployment limit of 1,000 weapons thus presents a mixed situation. It does reduce the size of the first-strike threat facing each state, should political relations deteriorate and fears of nuclear attack ever become realistic, compared to the 1,550 case. On the other hand, it also reduces the numbers of surviving and retaliating weapons for each state that provide the backbone of deterrence based on assured retaliation. The 1,000 case is thus a trade-off: additional political reassurance and more damage limitation in wartime, compared to the 1,550 deployment limit; on the other hand, deterrence might be less secure, especially extended deterrence for allies against attack or coercion, and leaders would have fewer post-attack options with their remaining forces.

Minimum deterrence

A second post-New START strategic nuclear arms control regime might be more ambitious than the reductions to a deployment limit of 1,000 weapons. Leaders might seek to reduce each of their peacetime numbers of operationally deployed weapons to several hundred instead of 1,000 weapons. Such a drastic step toward "minimum deterrence" would be welcomed by advocates of nuclear arms control, but how realistic would it be for US or Russian nuclear war planners or political leaders? In this analysis, we allocate a maximum number of 500 operationally deployed long-range nuclear weapons to each state. The numbers of second-strike surviving and retaliating warheads for the US and for Russia are calculated and displayed in Figure 12.3, immediately following.

The transition from an upper deployment limit of 1,000 to 500 deployed weapons is more significant than the step down from 1,550 to 1,000 warheads. In the smaller case of 500 deployed weapons, the assured retaliation mission can be accomplished but with little or no flexibility in targeting. Weapons will be allocated almost entirely if not entirely against cities and other economic and social assets. Few if any surviving weapons will be available for attacks on opposed nuclear forces, conventional forces, military command centers, or other targets.[8] The maintenance of a nuclear reserve force for post-attack bargaining, including escalation control and war termination, is all but impossible.

In addition, to live within the constraints of a "minimum deterrence" deployment of 500 or fewer weapons, Russia would have to realign fundamentally its current nuclear force structure and future modernization plans. At or below 500 weapons, silo-based intercontinental ballistic missiles (ICBMs) become serious

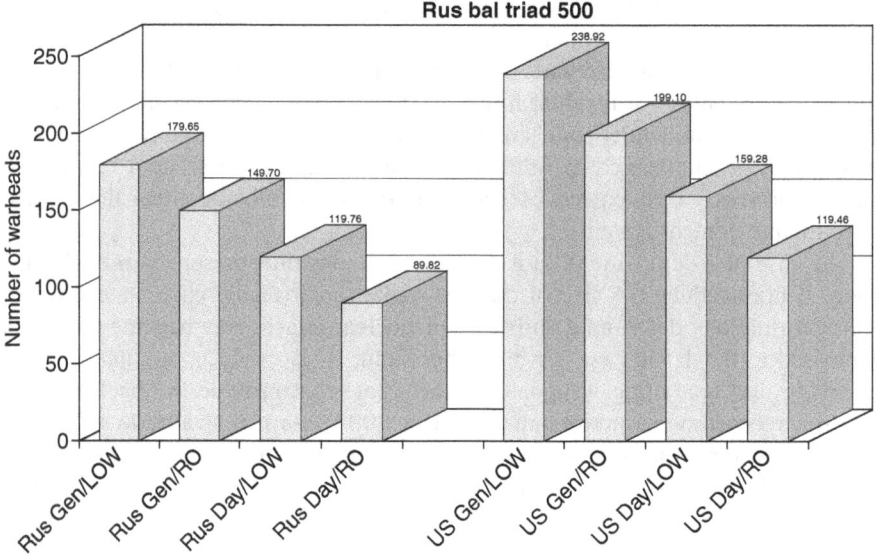

Figure 12.3 US–Russia surviving and retaliating warheads 500 deployment limit

liabilities because of their first-strike vulnerability compared to submarine launched ballistic missiles (SLBMs) or mobile ICBMs. Therefore, a Russian force downsized to 500 would have to relocate a larger proportion of its weapons on mobile land-based or sea-based missiles than it does now. Depending on the state of its economy and the competing priorities for modernizing its conventional military forces, Russia might be loath to abandon its ICBM-heavy strategic nuclear force structure, including some of its legacy silo-based ICBMs.

For the United States, a 500-warhead force would impose serious constraints on its ability to provide extended-deterrence commitments for its non-nuclear allies in Europe and Asia. Allies already feeling threatened by nuclear-armed regional neighbors, such as Japan or South Korea, might take more seriously the option of developing and deploying their own nuclear forces.[9] In addition, in order to preserve its preeminence in ballistic missile firing submarines and long-range bombers, the US might be forced to eliminate entirely the land-based missile segment of its nuclear "triad" or reduce its size to the point of triviality.[10]

Operating under a deployment limit of 500 weapons, Russia and the United States would also be challenged to maintain nuclear flexibility and resilience.[11] The options for equipping some portion of the nuclear retaliatory force with conventional warheads for prompt global strike missions would be restricted. So, too, might options for using nuclear weapons outside a US–Russian crisis, including the employment of precision, low-yield nuclear weapons against hardened targets such as nuclear or other WMD storage bunkers in states or under the control of non-state actors. Finally, there is the problem of missile defenses and their possible impact on the retaliatory deterrents of both the United States and Russia.

Missile defenses

The probable performance of missile defenses against offensive second-strike retaliation is unknown due to the uncertainties of current and future ballistic missile defense technologies.[12] Nevertheless it seems reasonable to assume that, *ceteris paribus*, the larger the offensive retaliatory force, the more challenging the problem is for the defense. In addition, missile defenses to protect populations, as opposed to retaliatory forces or other "hard" targets, are incredibly demanding because the arithmetic greatly favors the attacker. Even a small number of warheads penetrating a defense and aimed at population centers could create historically unprecedented destruction.

Thus, defenses of population centers have to be perfect or nearly perfect to be appealing to scientists or to deterrence theorists, but not necessarily to governments. Governments might reason that even imperfect defenses complicate the prospective first striker's attack plans, at least at the margin, and that in a crisis any hesitancy works in favor of the defender. This reasoning might be more compelling if defenses were deployed to protect retaliatory forces instead of populations. Two aspects of such a deployment might appeal to US planners. First, the technologies for reliable protection of land-based missiles have been available for years and no invention of new "wonder weapons" is required. Second, missile defenses for US retaliatory forces would obviously be intended to dilute an attacker's first strike, not his retaliatory second strike. Therefore, Russia's complaint, that US or NATO missile defenses are really intended to deter Russia and not Iran, would be less credible on that point. Table 12.1, immediately below, summarizes the generic approaches to US ICBM basing previously tried or proposed by the government or expert panels.

Another advantage of using missile defenses for retaliatory forces instead of cities is that the arithmetic is much more favorable to the defender, compared to the case of population defenses. Defenses of missile silos against first strikes, for example, need not perform perfectly in order to exert meaningful attrition against the attack. Even so-called simple-novel ground-based defenses or other available technologies could conceivably raise the "attack price" for destroying a silo from two to three or four warheads (or more), depending on accuracies and yields of the attackers.[13] If, for example, the US deploys some 400 Minuteman III ICBMs, this means that a first striker might have to plan for a salvo including some 1,200–1,600 warheads devoted to ICBMs alone. While this would not have been an impossible challenge for the Soviet Union during the Cold War, it would certainly stress the capacity of a New START-compliant nuclear first-strike force.

Would American or Russian ICBM defenses that incrementally raised the attack price against ICBM silos provide additional security worthy of the investment? In Figures 12.4 through 12.6, immediately following, we simulate the outcomes of nuclear exchanges at deployment levels of 1,550, 1,000 and 500

Table 12.1 Approaches to US ICBM basing and survivability

Generic approaches	Specific applications	Examples (systems/programs)
Hardening	• current or upgraded silos; • superhardened silos; • deep underground basing; • special configurations	• current silo program; • "Dense Pack"; • Mesa basing
Mobility	• ground mobile; • air mobile; • sea mobile	• road-mobile Minuteman; • small ICBM (SICBM); • rail-garrison MX ICBM; • cargo aircraft-ship; • small undersea mobile system (SUMS)
Deception/ concealment	• multiple aim points (MAPs); • concealed launch points	• MAPs; • hard trench; • deep underground basing
Defenses	• ballistic missile defense (BMD)	• low-altitude defense system (LoADS); • simple-novel silo defenses
Tactical responses	• launch under attack (LUA); • launch on warning (LOW)	LUA; LOW
Combinations	Hardening plus mobility	Hardened mobile launcher

Source: Lauren Caston et al., *The Future of the U.S. Intercontinental Ballistic Missile Force* (Santa Monica, CA: RAND Corporation, 2014), 37, adapted by author.

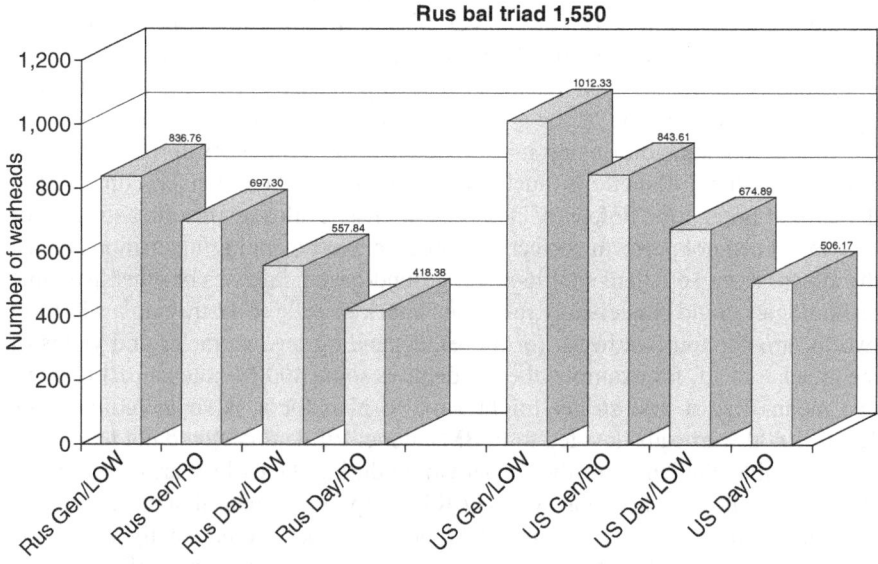

Figure 12.4 US–Russia surviving and retaliating warheads 1,550 deployment limit ICBM defenses

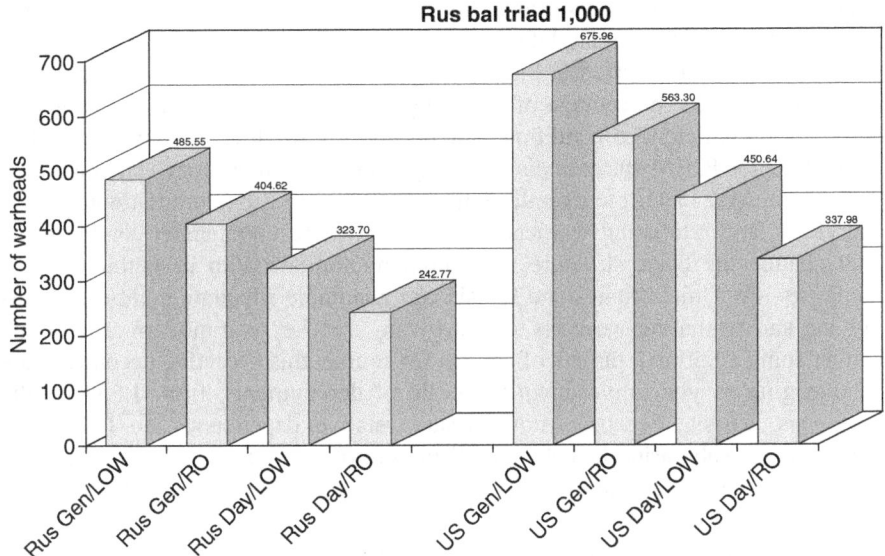

Figure 12.5 US–Russia surviving and retaliating warheads 1,000 deployment limit ICBM defenses

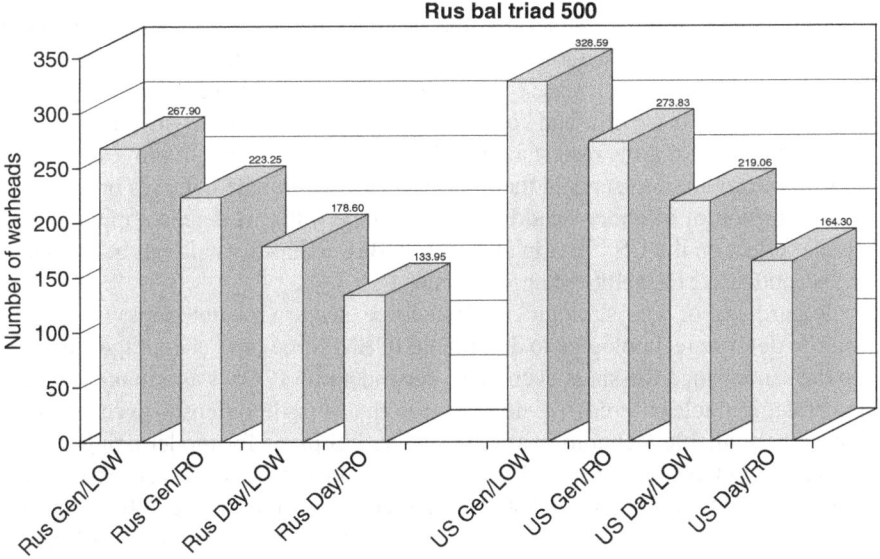

Figure 12.6 US–Russia surviving and retaliating warheads 500 deployment limit ICBM defenses

warheads for the United States and for Russia, each having deployed ICBM defenses that increase the survivability of silo-based missiles to that of mobile land-based missiles without defenses.

The results summarized in Figures 12.4 through 12.6 show that silo defenses could increase the percentage of surviving and retaliating ICBM warheads, relative to the undefended condition. The question is whether the improvement in outcomes for ICBM survivors is meaningful in terms of strategy. The picture is mixed. On one hand, the overall numbers of US and Russian additional ICBM surviving and retaliating warheads, compared to the undefended condition, do not change the basic structure of assured retaliation. With or without ICBM defenses, the United States and Russia can guarantee adequate numbers of surviving and retaliating weapons to destroy any attacker as a modern society and cover some additional targets of choice. Of course, this objective becomes more challenging as you move down the scale of deployments, from 1,550 to 500 weapons. Russia, because of its greater relative dependency on ICBMs as opposed to submarine-launched ballistic missiles (SLBMs), gains relatively more than the United States does under the assumption of technologically symmetrical silo defense deployments.

On the other hand, from a military and deterrence standpoint, there are two potential benefits from ICBM defenses compared to the undefended condition. First, any attacker could not know the exact performance of ICBM defenses under crisis or wartime conditions: even the defender would be estimating based on tests and simulations. These unknown parameters of missile defense performance "under fire" might complicate the attacker's first-strike confidence. Second, the availability of missile defenses could allow US or Russian leaders to feel less pressure to "use them or lose them" and increase confidence against a mistaken decision for preemption. Apart from these pros and cons of deterrence and nuclear-crisis stability, there is also the issue of arms race stability. US ICBM defenses might provoke Russian or Chinese countermeasures in the form of their own missile defenses or offsetting modernization of offences—and Russian or Chinese ICBM defenses might have a similar effect on the US.[14] But in all cases, ICBM defenses would not be a threat to the second-strike capability of another state.

Regardless of US strategic nuclear force size, why not deploy available missile defense technologies to defend the ICBM force (and encourage Russia to do the same, since Russia is even more dependent on ICBMs as a makeweight of its strategic nuclear forces)? One reason is that missile defenses even based on currently available technology are expensive. US plans for modernizing nuclear forces, command systems, and infrastructure, including weapons laboratories, are already bumping up against the budget sequestration and other expected fiscal constraints.[15] Another reason is that two of the three components of the US nuclear triad, submarines and bombers, are presumptively survivable without missile defenses, provided (especially in the case of bombers) sufficient warning is available.

But a third reason for lack of interest in missile defense for ICBMs is more controversial from an arms control perspective. ICBMs are the quick-reaction component of the US and Russian nuclear triads. Although SLBMs can also be tasked for prompt launch missions, their uniqueness lies in their unmatched survivability (at least for now). ICBMs also do not have to move to another location before firing after having received duly authorized launch commands, as ballistic missile firing submarines (SSBNs) sometimes do. In addition, although the command, control, and communications (C3) systems for ballistic missile submarines are reportedly as reliable as those for strategic land-based missiles, the latter are not quite as complicated.[16] On the other hand, some arms control experts maintain that the United States and Russia maintain too many ICBM warheads on ready alert for prompt launch, creating a "hair trigger" problem during a prospective nuclear crisis.[17] And other arms control experts argue that the Cold War history of strategic missile defenses, whether deployed by the US or by the Soviet Union, was that they generated offsetting changes in the other side's force modernization and nuclear targeting plans—including specific plans for the suppression of missile defenses.[18]

What about the possibility that the US and Russia might deploy both ICBM silo defenses against first strikes and wider area defenses for populations against retaliatory strikes? These two missions for anti-missile defenses might seem contradictory according to deterrence logic, but politics might collude with improved technology to create a "both, and" as well as an "either, or" option. Figures 12.7 through 12.9, immediately below, summarize the outcomes of nuclear force exchanges at various deployment levels, under the assumptions of

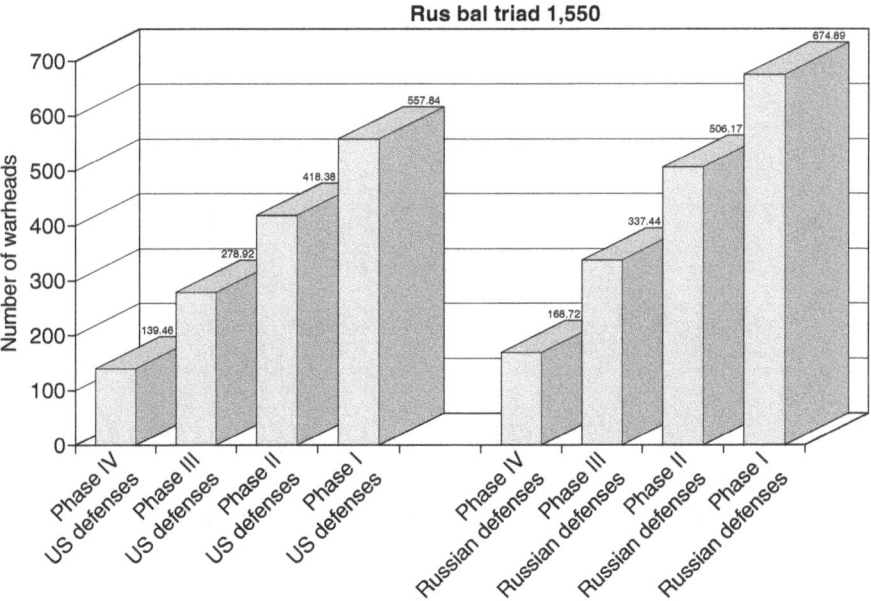

Figure 12.7 US–Russia surviving and retaliating warheads 1,550 deployment limit ICBM and population defenses

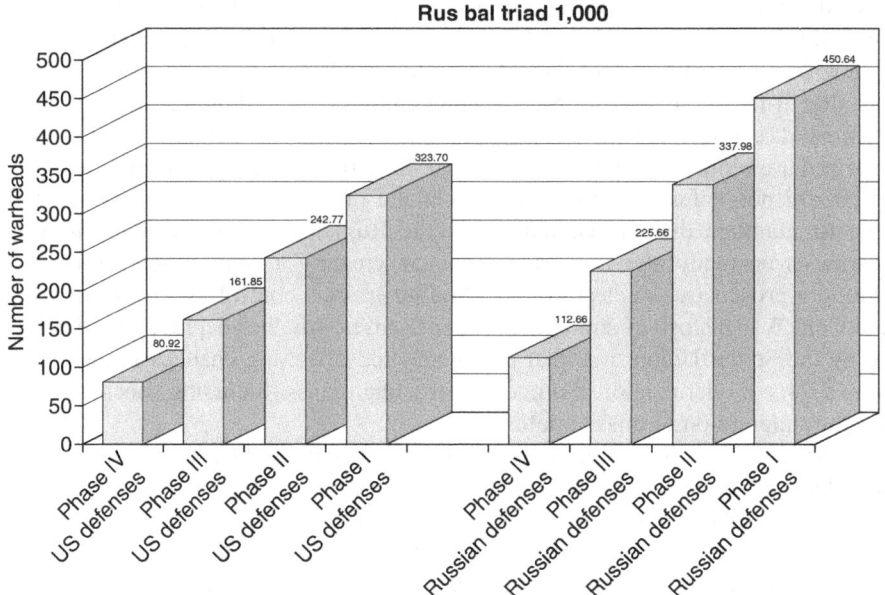

Figure 12.8 US–Russia surviving and retaliating warheads 1,000 deployment limit ICBM and population defenses

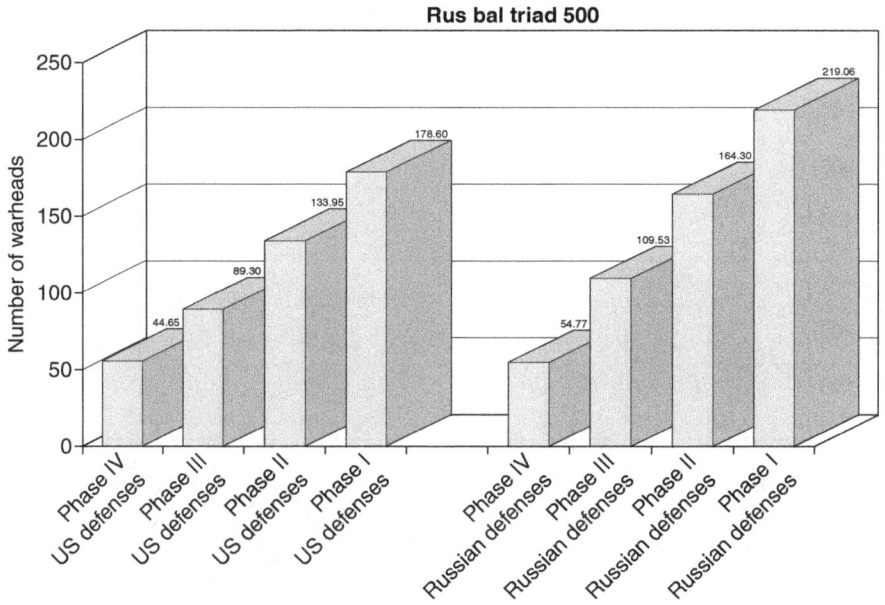

Figure 12.9 US–Russia surviving and retaliating warheads 500 deployment limit ICBM and population defenses

ICBM silo defenses (as previously discussed) and population defenses for both states. Since the actual performances of future population defenses are even more conjectural than those for silo defenses, population defenses are given a sliding scale of capabilities: Phase I defenses intercept or otherwise deflect at least 20 percent of second-strike retaliating warheads; Phase II, at least 40 percent; Phase III, at least 60 percent; and Phase IV, at least 80 percent.

The outcomes summarized in Figures 12.7 through 12.9 suggest several inferences and possibilities. First, one gets more bang for the buck by adding increments of silo defenses against first strikes, compared to increments of population defenses against retaliatory strikes, unless or until defenses are perfect or nearly so. This finding reflects the greater technology and strategy challenges faced by population defenses, compared to silo defenses. Second, Russia might blanch at the "defenses abundant" scenario because Russia could fear that a competition in deploying population defenses, compared to silo defenses, would strain its resources for nuclear force modernization. Third, no combination of silo and/or population defenses could remove either the US or Russia away from its mutual deterrence and reciprocal nuclear hostage situation, given current or imminently foreseeable technologies.

Nonproliferation

The spread of nuclear weapons is regarded as an imminent danger by some and as an inevitable or acceptable outcome by others.[19] Regardless the inclinations of theorists, few in the US or allied NATO situation rooms are prepared to applaud or encourage nuclear weapons spread. The reason for this practical stance of proliferation skepticism is that the most likely candidates for near-term nuclear acquisition are potentially disruptive actors (Iran) or states located in hothouses of political and military rivalry (the Middle East and East and South Asia). There is also the risk that incautious or mischievous state actors may share nuclear material, weapons, or know-how with terrorists. A third reason for proliferation pessimism is that new nuclear weapons states may have irresponsible political leadership or unaccountable civil–military relations that make control over nuclear forces suspect or dangerous. (For example, imagine a future scenario in which North Korea's political system melts down and its custody and use controls over nuclear forces are insecure.)

On the other hand, states that feel threatened by adversaries with superior conventional military power, or by the possibility of nuclear coercion, may turn to nuclear weapons as a deterrent or military checkmate. Pakistan is an example of a state that continues to build its nuclear arsenal as part of its political and military rivalry with India. India, in turn, feels the need to stay ahead of Pakistan and to compete with a rising China. China's waxing power and political assertiveness in the East and South China Seas, and North Korea's exuberant malice, pose potential challenges to others in the region, including Japan and South Korea.[20]

The motivations for states feeling threatened by the prospect of nuclear coercion or attack to want nuclear arsenals for deterrence may be understandable, but not necessarily commendable. Much depends on the quality of political leadership and on the stability of civil–military relations in future states that acquire nuclear arsenals. As to the first, political leadership, those holding the reins of power in new nuclear weapons states will have a steep learning curve with respect to the political oversight of nuclear forces and the management of nuclear force operations. As to the second, the stability of civil–military relations, this has two aspects: first, the authority and credibility of the civilian government leadership and bureaucracy; and, second, the reliability of delegations of authority to the armed forces and nuclear use controls during periods of crisis and stress as well as under normal peacetime conditions.

As the US and Soviet experiences in the Cuban missile crisis and in other Cold War and later incidents have demonstrated, none of the items listed in the preceding paragraph can be taken for granted, even in the case of mature nuclear powers.[21] This vulnerability of domestic politics and civil-military relations to deficiencies in nuclear crisis management becomes especially significant when political systems from different cultures, historical experiences, and philosophical traditions come into conflict. The very idea of "deterrence" as a methodology based on rational decision making presumes a certain commonality in states' shared perceptions of costs and benefits. But these costs and benefits can be reckoned in very different ways by different regimes.

Regimes afflicted with a strong sense of domestic insecurity or holding extreme nationalistic ambitions might take the challenge of brinkmanship from a different, and more dangerous, standpoint, compared to states that are domestically secure and ideologically moderate. In addition, as Lawrence Freedman has explained, deterrence logic often depended upon the exploitation of uncertainty and the willingness to tolerate a degree of risk that a process of escalation would, in fact, get out of control:

> The uncertainty would grow as a crisis turned into a limited conflict and then moved toward general war, getting out of hand "by degrees." Skillful tactics would exploit this fact, not shrink from it. The assumption was that it was worth letting a "situation get somewhat out of hand" because the opponent would find such circumstances intolerable. Deterrence was possible because of a situation in which terrible things *might* happen (which was credible because of human irrationality) rather than a specific threat to do those things (which was incredible because of human rationality).[22]

In addition to the possible disparities in states' risk acceptance under conditions of extreme uncertainty, the problem of nuclear crisis management could also be exacerbated by developments in technology. States' growing interest in cyberwar in the twenty-first century might seem to be a hallmark example of nondestructive warfare, and therefore, partaking of an entirely different universe than that occupied by nuclear weapons and deterrence. This impression of totally distinct nuclear and cyber universes without any behavioral intersection is misleading. Deterrence is a psychological process and is highly dependent upon perceptions and expectations modified, more or less, by available evidence, including intelligence. In addition, the successful practice of nuclear crisis management requires the technical means and the mindsets of leaders for reliable communication of decisions and intentions. The introduction of cyber attacks before or during a nuclear crisis could contribute to misperceptions and misleading communications, or cause disruptions in information networks and systems for C4ISR (command, control, communications, computers, intelligence, surveillance, and reconnaissance). For example, cyber attacks on satellite networks could reduce the amount of reliable information one side needs to reassure itself that the other side was not, in fact, engaged in nuclear preemption.[23]

Finally, the issue of cyber war, like the issue of missile defense, is a reminder of the fact that the major nuclear powers are especially capable and therefore responsible for the control of nuclear weapons spread. Their mature forces and command-control systems have benefited from "lessons learned" across decades, and, in the US and Russian cases, from the very dawn of the nuclear age. Nuclear war in the future is not going to be started or prevented by autonomous expert systems or robot commanders. But future nuclear crisis management and conventional force operations will take place in a digital, not an analog world, and the politico-military technology matrix for deterrence and defense policy will inevitably differ from that of the Cold War. Instead of being frustrated by information scarcity, future leaders will be overwhelmed by digital confusion and pandemic information overload. The challenge will be to sort it all out in good time: within the mission envelope for war avoidance, or victory.

Conclusion

The United States and Russia have a need and a responsibility to continue dialogue on nuclear arms reductions, regardless of the ups and downs of controversy on other issues. The apparent violation by Russia of the Intermediate Nuclear Forces (INF) treaty, by having tested a prohibited ground-launched cruise missile, casts another shadow over nuclear arms reduction efforts.[24] But, even during the most frustrating periods of Cold War politics, the Soviets and Americans negotiated strategic nuclear arms control agreements from which both sides benefited. Dramatic shifts in political moods affecting the probability of successful arms control can follow changes of leaders in the United States or in Russia.[25] US and Russian willingness to limit the sizes and capabilities of their own nuclear arsenals does not guarantee that they will succeed in leading international security management on nonproliferation. But, if Russian and American leadership on the prevention of nuclear weapons spread is not a sufficient condition for success, it is certainly a necessary condition. And both states have a common interest in keeping nuclear weapons and materials out of the hands of irresponsible regimes or terrorists.

On the other hand, US–Russian strategic nuclear arms reductions must take into account the requirements of both states' policies and requirements for nuclear force operations. In that respect, reductions in the numbers of American and Russian operationally deployed weapons on intercontinental launchers, below the New START maximum of 1,550 warheads on 700 launchers, are likely to be only incremental and not a drastic departure from New START levels. Analysis suggests that reductions to 1,000 or so peacetime deployed weapons would leave policymakers and military planners in Washington and Moscow with the nuclear flexibility and resilience they require for their assigned missions and for unforeseen "Black Swan" circumstances. On the other hand, more ambitious reductions to "minimum deterrence" levels of several hundred weapons could reduce nuclear flexibility and resilience to unacceptable levels, especially for US extended deterrence requirements and for Russia's need for nuclear compensation against conventional force deficiencies.

Symmetrical nuclear arms reductions as between the United States and Russia may no longer have symmetrical effects, as assumed to be the case during the Cold War. According to Keith Darden and Timofei Bordachev, the US and Russia should seek not only strategic stability based on mutual deterrence, but also strategic compatibility, allowing for differing but compatible security portfolios.[26] For example, future US modernization might emphasize anti-missile defenses, while Russian approaches would favor offenses. Granted, US and NATO missile defenses have sparked Russian protests over fears that Russia's nuclear deterrent will be compromised. But missile defenses for nuclear retaliatory forces, especially ICBMs, arguably complicate a first strike, not a guaranteed second strike. In addition, silo defenses contribute to first-strike stability, and also to crisis stability, because they protect the most first-strike vulnerable portion of the US and Russian strategic nuclear triads. On the other hand, a viable alternative to undefended silo basing will have to be cost-effective and lacking in toxic political baggage that has weighed down some past programs.[27]

Notes

1 For an interesting assessment of this conflict, see Lawrence Freedman, "Ukraine and the Art of Limited War," warontherocks.com, http://warontherocks.com/2014/10/ukraine-and-the-art-of-limited-war/html.
2 "Putin Threatens Nuclear War over Ukraine," *The Daily Beast*, August 31, 2014, www.thedailybeast.com/articles/2014/08/31/putin-threatens-nuclear-war-over-ukraine.html. See also "Russia's Strategic Nuclear Forces to Hold Major Exercise This Month," Reuters, September 3, 2014, http://uk.reuters.com/article/2014/09/03/uk-ukraine-crisis-russia-exercises-idUKKBNOGYOH620140903.
3 Jason Groves, "It's Better Not to Mess with Russia: Putin's Nuclear Warning to West on Ukraine," *The Daily Mail Online*, August 29, 2014, www.dailymail.co.uk/news/article-2737526/Putin-s-plans-revealed.
4 Vladimir Isachenkov, "Putin Promises New Weapons to Fend Western Threats," Associated Press, September 10, 2014, http://hosted.ap.org/dynamic/stories/E/EU_Russia_Putin?SITE=AP&S.
5 Vladimir Filonov, "Russian General Calls for Preemptive Nuclear Strike Doctrine Against NATO," *Moscow Times*, September 3, 2014, www.themoscowtimes.com/business/article/russian-general-calls-for-preemptive-nuclear-strike-doctrine-against-nato/506370.html.

6 US Department of State, "Treaty between the United States of America and the Russian Federation on Measures for the Further Reduction and Limitation of Strategic Offensive Arms," April 8, 2010, www.state.gov/documents/organization/140035.pdf.

7 Grateful acknowledgment is made to Dr. James J. Tritten for use of a model originally developed by him in portions of this study. Tritten is not responsible for its use here nor for any arguments or opinions in this study.

8 See US Department of Defense, *Report on Nuclear Employment Strategy of the United States Specified in Section 491 of 10 U.S.C.*, June 12, 2013, for a discussion of principles guiding the use of US nuclear forces. This document notes that new guidance requires the United States "to maintain significant counterforce capabilities against potential adversaries" and emphasizes that current policy "does not rely on a 'counter-value' or 'minimum deterrence' strategy," p. 4.

9 For example, see Doug Bandow, "Maybe U.S. Should Defend South Korea by Letting it Develop Nuclear Weapons," Cato Institute, August 11, 2014, www.cato.org/blog/maybe-us-should-defend-south-korea-letting-it-develop-nuclear-weapons.

10 For expert assessment of alternative futures for the US strategic land-based missile force, see Lauren Caston, Robert S. Leonard, Christopher A. Mouton, Chad J. R. Ohlandt, S. Craig Moore, Raymond E. Conley, and Glenn Buchan, *The Future of the U.S. Intercontinental Ballistic Missile Force* (Santa Monica, CA: RAND Corporation, 2014).

11 On the requirements for nuclear flexibility and resilience, see Keith B. Payne, Study Director, and John S. Foster Jr., Chairman, Senior Review Group, *Nuclear Force Adaptability for Deterrence and Assurance: A Prudent Alternative to Minimum Deterrence* (Fairfax, VA: National Institute Press, 2014). Arguments in favor of minimum deterrence appear in James Wood Forsyth Jr., B. Chance Saltzman, and Gary Schaub Jr., "Minimum Deterrence and Its Critics," *Strategic Studies Quarterly* 4 (Winter 2010): 3–12; and Forsyth, Saltzman, and Schaub, "Remembrance of Things Past: The Enduring Value of Nuclear Weapons," *Strategic Studies Quarterly* 1 (Spring 2010): 74–89. See also Bruce Blair, Victor Esin, Matthew McKinzie, Valery Yarynich, and Pavel Zolotarev, "One Hundred Nuclear Wars: Stable Deterrence between the United States and Russia at Reduced Nuclear Force Levels Off Alert in the Presence of Limited Missile Defenses," *Science and Global Security* 19 (2011): 167–194, DOI: 10.1080/08929882.2011.616127.

12 For example, see Committee on an Assessment of Concepts and Systems for U.S. Boost-Phase Missile Defense in Comparison to Other Alternatives, *Making Sense of Ballistic Missile Defense: An Assessment of Concepts and Systems for U.S. Boost-Phase Missile Defense in Comparison to Other Alternatives* (Washington, DC: National Research Council, National Academy of Sciences, National Academies Press, 2012), prepublication copy, www.nap.edu. See also George N. Lewis and Theodore A. Postol, "A Flawed and Dangerous U.S. Missile Defense Plan," *Arms Control Today*, May 2010, www.armscontrol.org/act/2010_05/Lewis-Postol.

13 Ashton B. Carter, "BMD Applications: Performances and Limitations," Ch. 4 in Carter and David N. Schwartz, eds., *Ballistic Missile Defense* (Washington, DC: Brookings Institution, 1984), 98–181, esp. 126–128.

14 For commentary on Russian and Chinese missile and space defense efforts, see "Russia Plans $55.3 Bln Expenditure on Aerospace Defense by 2020," RIA Novosti, February 28, 2014, http://en.ria.ru/news/20140228/187971313/Russia-Plans-553Bln-Expenditure-On-Aerospace-Defense-by-2020.html; Michaela Dodge, "U.S. Missile Defense Policy After Russia's Actions in Ukraine," Heritage Foundation *Issue Brief* No. 4177, March 21, 2014, www.heritage.org/research/reports/2014/03/us-missile-defense-policy-after-russia's-actions-in-ukraine.html; and Ting Shi, "China Says Third Missile-Defense Test in Four Years Successful," Bloomberg.com, July 24, 2014, www.bloomberg.com/news/print/2014-07-24/china-says-third-missile-defense-test-in-four-years-successful.html. See also David S. Forman, "Deterrence with

China: Avoiding Nuclear Miscalculation," *Joint Forces Quarterly* 75 (4th Quarter 2014): 34–42.

15 See Jon B. Wolfsthal, Jeffrey Lewis, and Marc Quint, "The Trillion Dollar Triad: U.S. Strategic Modernization Over the Next Thirty Years," James Martin Center for Non-proliferation Studies, January 2014; and Hans M. Kristensen and Robert S. Norris, US Nuclear Forces 2014, *Bulletin of the Atomic Scientists* 1 (2014): 85–93, esp. 88, http://bos.sagepub.com.

16 Bruce G. Blair, *Strategic Command and Control: Redefining the Nuclear Threat* (Washington, DC: Brookings Institution, 1985), Ch. 7–8 and passim.

17 Global Zero Nuclear Policy Commission, *Report: Modernizing U.S. Nuclear Strategy, Force Structure and Posture* (Washington, DC: Global Zero, May 2012), www.globalzero.org.

18 Hans M. Kristensen, Matthew G. McKinzie, and Robert S. Norris, "The Protection Paradox," *Bulletin of the Atomic Scientists* (March/April 2004): 68–77.

19 For a typology of optimists and pessimists on nuclear proliferation, see Peter D. Feaver, "Nuclear Command and Control in Crisis: Old Lessons from New History," Ch. 7 in Henry D. Sokolski and Bruno Tertrais, eds., *Nuclear Weapons Security Crises: What Does History Teach?* (Carlisle, PA: Strategic Studies Institute and US Army War College Press, July 2013), 205–225. See also Scott D. Sagan and Kenneth N. Waltz, *The Spread of Nuclear Weapons: A Debate* (New York: W. W. Norton, 1995).

20 Kang Seung-woo, "NK Could Play Nuclear Option," *Korea Times*, August 12, 2014, http://m.koreatimes.co.kr/phone/news/views.jsp?req_newsidx=162687. See also "U.S. Prepared to Increase Pressure on North Korea: Kerry," Channel NewsAsia, August 12, 2014, www.channelnewsasia.com/news/asiapacific/us-prepared-to-increase/1308 220.html.

21 For example, see the case studies in Sokolski and Tertrais, eds., *Nuclear Weapons Security Crises: What Does History Teach?*

22 Lawrence Freedman, *Strategy: A History* (New York: Oxford University Press, 2013), 165. This commentary is in the context of reviewing work by Thomas C. Schelling, especially Schelling's *The Strategy of Conflict* (Cambridge, MA: Harvard University Press, 1960) and *Arms and Influence* (New Haven, CT: Yale University Press, 1966).

23 Cyber attacks are not necessarily decisive in themselves, apart from kinetic attacks, and the consequences of nuclear war and cyber attacks are, for obvious reasons, very different. See Colin S. Gray, *Making Strategic Sense of Cyber Power: Why the Sky Is Not Falling* (Carlisle, PA: Strategic Studies Institute, US Army War College, April 2013). See also Thomas M. Chen, *An Assessment of the Department of Defense Strategy for Operating in Cyberspace* (Carlisle, PA: Strategic Studies Institute, US Army War College Press, September 2013).

24 Michael R. Gordon, "U.S. Says Russia Tested Cruise Missile, Violating Treaty," *New York Times*, July 28, 2014, www.nytimes.com/2014/07/29/world/europe/us-says-russia-tested-cruise-missile-violating-treaty.

25 See George Friedman, "Can Putin Survive?" Stratfor.com, July 21, 2014, in *Johnson's Russia List* 2014 – #159 – July 22, 2014, davidjohnson@starpower.net.

26 Keith Darden and Timofei Bordachev, "The Sword and the Shield: Toward U.S.–Russian Strategic Compatibility," Working Group Paper 4, Working Group on the Future of U.S.–Russian Relations, Davis Center for Russian and Eurasian Studies, Harvard University, September 2014, http://us-russiafuture.org/publications.

27 This includes "public interface" safety and security issues such as transportation of nuclear or other materials. See Caston et al., *The Future of the U.S. Intercontinental Ballistic Missile Force*, 34–35.

Bibliography

Bandow, Doug, "Maybe U.S. Should Defend South Korea by Letting it Develop Nuclear Weapons," Cato Institute, August 11, 2014, www.cato.org/blog/maybe-us-should-defend-south-korea-letting-it-develop-nuclear-weapons.

Blair, Bruce G., *Strategic Command and Control: Redefining the Nuclear Threat* (Washington, DC: Brookings Institution, 1985).

Blair, Bruce, Esin, Victor, McKinzie, Matthew, Yarynich, Valery, and Zolotarev, Pavel, "One Hundred Nuclear Wars: Stable Deterrence between the United States and Russia at Reduced Nuclear Force Levels Off Alert in the Presence of Limited Missile Defenses," *Science and Global Security* 19 (2011), DOI: 10.1080/08929882.2011.616127.

Carter, Ashton B., "BMD Applications: Performances and Limitations," Ch. 4 in Carter and David N. Schwartz, eds., *Ballistic Missile Defense* (Washington, DC: Brookings Institution, 1984).

Caston, Lauren, Leonard, Robert S., Mouton, Christopher A., Ohlandt, Chad J. R., Moore, S. Craig, Conley, Raymond E., and Buchan, Glenn, *The Future of the U.S. Intercontinental Ballistic Missile Force* (Santa Monica, CA: RAND Corporation, 2014).

Chen, Thomas M., *An Assessment of the Department of Defense Strategy for Operating in Cyberspace* (Carlisle, PA: Strategic Studies Institute, US Army War College Press, September 2013).

Committee on an Assessment of Concepts and Systems for U.S. Boost-Phase Missile Defense in Comparison to Other Alternatives, *Making Sense of Ballistic Missile Defense: An Assessment of Concepts and Systems for U.S. Boost-Phase Missile Defense in Comparison to Other Alternatives* (Washington, DC: National Research Council, National Academy of Sciences, National Academies Press, 2012), prepublication copy, www.nap.edu.

Darden, Keith, and Bordachev, Timofei, "The Sword and the Shield: Toward U.S.–Russian Strategic Compatibility," Working Group Paper 4, Working Group on the Future of U.S.–Russian Relations, Davis Center for Russian and Eurasian Studies, Harvard University, September 2014, http://us-russiafuture.org/publications.

Dodge, Michaela, "U.S. Missile Defense Policy After Russia's Actions in Ukraine," Heritage Foundation *Issue Brief* No. 4177, March 21, 2014, www.heritage.org/research/reports/2014/03/us-missile-defense-policy-after-russia's-actions-in-ukraine.html.

Feaver, Peter D., "Nuclear Command and Control in Crisis: Old Lessons from New History," Ch. 7 in Henry D. Sokolski and Bruno Tertrais, eds., *Nuclear Weapons Security Crises: What Does History Teach?* (Carlisle, PA: Strategic Studies Institute and U.S. Army War College Press, July 2013).

Filonov, Vladimir, "Russian General Calls for Preemptive Nuclear Strike Doctrine Against NATO," *Moscow Times*, September 3, 2014, www.themoscowtimes.com/business/article/russian-general-calls-for-preemptive-nuclear-strike-doctrine-against-nato/506370.html.

Forman, David S., "Deterrence with China: Avoiding Nuclear Miscalculation," *Joint Forces Quarterly* 75 (4th Quarter 2014).

Forsyth Jr., James Wood, Saltzman, B. Chance and Schaub Jr., Gary, "Minimum Deterrence and Its Critics," *Strategic Studies Quarterly* 4 (Winter 2010).

Forsyth Jr., James Wood, Saltzman, B. Chance and Schaub Jr., Gary, "Remembrance of Things Past: The Enduring Value of Nuclear Weapons," *Strategic Studies Quarterly* 1 (Spring 2010).

Freedman, Lawrence, *Strategy: A History* (New York: Oxford University Press, 2013).

Freedman, Lawrence, "Ukraine and the Art of Limited War," warontherocks.com, http://warontherocks.com/2014/10/ukraine-and-the-art-of-limited-war/html.

Friedman, George, "Can Putin Survive?" Stratfor.com, July 21, 2014, in *Johnson's Russia List* 2014 – #159 – July 22, 2014, davidjohnson@starpower.net.

Global Zero Nuclear Policy Commission, *Report: Modernizing U.S. Nuclear Strategy, Force Structure and Posture* (Washington, DC: Global Zero, May 2012), www.globalzero.org.

Gordon, Michael R., "U.S. Says Russia Tested Cruise Missile, Violating Treaty," *New York Times*, July 28, 2014, www.nytimes.com/2014/07/29/world/europe/us-says-russia-tested-cruise-missile-violating-treaty.

Gray, Colin S., *Making Strategic Sense of Cyber Power: Why the Sky Is Not Falling* (Carlisle, PA: Strategic Studies Institute, US Army War College, April 2013).

Groves, Jason, "It's Better Not to Mess with Russia: Putin's Nuclear Warning to West on Ukraine," *The Daily Mail Online*, August 29, 2014, www.dailymail.co.uk/news/article-2737526/Putin-s-plans-revealed.

Isachenkov, Vladimir, "Putin Promises New Weapons to Fend Western Threats," Associated Press, September 10, 2014, http://hosted.ap.org/dynamic/stories/E/EU_Russia_Putin?SITE=AP&S.

Kristensen, Hans M., McKinzie, Matthew G., and Norris, Robert S., "The Protection Paradox," *Bulletin of the Atomic Scientists* (March/April 2004).

Kristensen, Hans M., and Norris, Robert S., "US Nuclear Forces 2014," *Bulletin of the Atomic Scientists* 1 (2014), http://bos.sagepub.com.

Lewis, George N., and Postol, Theodore A., "A Flawed and Dangerous U.S. Missile Defense Plan," *Arms Control Today*, May 2010, www.armscontrol.org/act/2010_05/Lewis-Postol.

Payne, Keith B., Study Director, and Foster Jr., John S., Chairman, Senior Review Group, *Nuclear Force Adaptability for Deterrence and Assurance: A Prudent Alternative to Minimum Deterrence* (Fairfax, VA: National Institute Press, 2014).

"Putin Threatens Nuclear War over Ukraine," *The Daily Beast*, August 31, 2014, www.thedailybeast.com/articles/2014/08/31/putin-threatens-nuclear-war-over-ukraine.html.

"Russia Plans $55.3 Bln Expenditure on Aerospace Defense by 2020," RIA Novosti, February 28, 2014, http://en.ria.ru/news/20140228/187971313/Russia-Plans-553Bln-Expenditure-On-Aerospace-Defense-by-2020.html.

"Russia's Strategic Nuclear Forces to Hold Major Exercise This Month," Reuters, September 3, 2014, http://uk.reuters.com/article/2014/09/03/uk-ukraine-crisis-russia-exercises-idUKKBNOGYOH620140903.

Sagan, Scott D., and Waltz, Kenneth N., *The Spread of Nuclear Weapons: A Debate* (New York: W. W. Norton, 1995).

Seung-woo, Kang, "NK Could Play Nuclear Option," *Korea Times*, August 12, 2014, http://m.koreatimes.co.kr/phone/news/views.jsp?req_newsidx=162687.

Schelling, Thomas C., *Arms and Influence* (New Haven, CT: Yale University Press, 1966).

Schelling, Thomas C., *The Strategy of Conflict* (Cambridge, MA: Harvard University Press, 1960).

Shi, Ting, "China Says Third Missile-Defense Test in Four Years Successful," Bloomberg.com, July 24, 2014, www.bloomberg.com/news/print/2014-07-24/china-says-third-missile-defense-test-in-four-years-successful.html.

Sokolski, Henry D., and Tertrais, Bruno, eds., *Nuclear Weapons Security Crises: What Does History Teach?* (Carlisle, PA: Strategic Studies Institute and US Army War College Press, July 2013).

US Department of Defense, *Report on Nuclear Employment Strategy of the United States Specified in Section 491 of 10 U.S.C.*, June 12, 2013.

US Department of State, "Treaty between the United States of America and the Russian Federation on Measures for the Further Reduction and Limitation of Strategic Offensive Arms," April 8, 2010, www.state.gov/documents/organization/140035.pdf.

"U.S. Prepared to Increase Pressure on North Korea: Kerry," Channel NewsAsia, August 12, 2014, www.channelnewsasia.com/news/asiapacific/us-prepared-to-increase/1308220.html.

Wolfsthal, Jon B., Lewis, Jeffrey, and Quint, Marc, "The Trillion Dollar Triad: U.S. Strategic Modernization Over the Next Thirty Years," James Martin Center for Nonproliferation Studies, January 2014.

Index